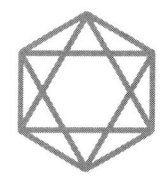

원서읽기를 위한
수학용어사전
수정판

원서읽기를 위한

수학용어사전

초판발행 2018년 08월 31일
수정발행 2022년 10월 01일

지 은 이 조민형
펴 낸 곳 지오북스
주   소 서울시 중구 퇴계로41길 39, 3층 302호(정암프라자)
등   록 2016년 3월 7일 제395-2016-000014호
전   화 02)736-0640 | 팩스 02)371-0706
이 메 일 emotion-books@naver.com
홈페이지 www.geobooks.co.kr

ISBN 979-11-91346-41-1
값 19,000원

이 책은 저작권법으로 보호받는 저작물입니다.
이 책의 내용을 전부 또는 일부를 무단으로 전재하거나 복제할 수 없습니다.
파본이나 잘못된 책은 바꿔드립니다.

머리말

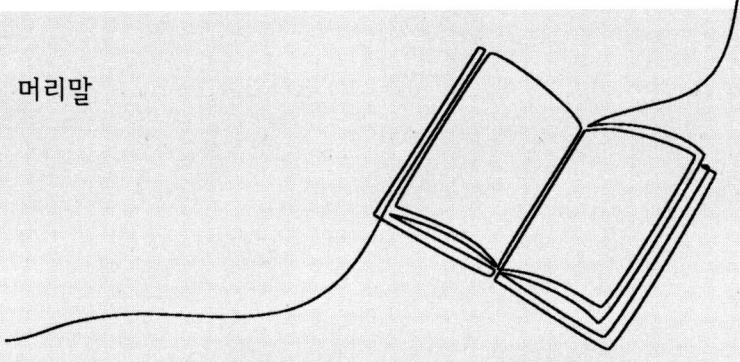

이 책은 1983년 가을학기에 미국에서 대학원 첫 학기 조교를 하면서 학생들에게 영어로 설명할 때 정리한 자료를 시작으로, 미국의 초중등교과서에서 나오는 수학용어를 수집하여 정리하고, 귀국하여 금오공과대학교에서 수학 강의 준비하면서 모아둔 상당한 분량의 자료 들을 정리하면서 얻어진 내용을 모은 것이다. 수집한 자료에 비해 책의 수록할 내용의 제한 때문에 많은 부분은 별도의 파일로 남게 된 점을 아쉽게 생각한다.

영어 용어를 중심으로 수학을 원서로 학습하게 될 학생들에게 도움이 되도록 자료를 수집하고 편집하였다. 영어 용어는 보편적으로 많이 쓰이는 단어를 중심으로 작성하였으며, 유사한 표현과 연결이 되도록 구성하였다.

영어 이외에 고등학교와 대학교에서 제2외국어로 학습한 프랑스어 얕팍한 지식과, 박사학위 논문을 작성하면서 지도교수께서 알려주신 러시아 논문검색 경험에 따라 간단한 러시아어 발음과 인명, 그리고 미국 New Mexico 주와 Texas 주에서 몇 년간 생활한 덕분에 얻어진 스페인어에 대한 자료도 첨가하였다. 물론이 자료만 가지고 제2외국어로 된 수학원서를 이해하기는 힘들기 때문에 해당언어의 초급과정 학습을 전제로 하였다.

본인의 능력 부족은 많은 분들의 도움으로 줄여나갈 수 있었다. 미국 세인트루이스에서 학생을 지도하고 있는 폴란드 친구 V. Golik 교수와 New Mexico State University 에서 수학, 산업공학과 경영학과에서 전공 강의를 하시는 등 넓은 지식을 가졌으며 박사논문 지도교수인 M. Maher 교수, 그리고 러시아어로 된 논문을 한국어로 직접 번역하여 박사학위 논문 작성에 도움을 주셨으며 중국어 수학서적에 나오는 수학적인 표현을 지도해 주셨던 하얼빈공업대학의 고 李容泉 교수(1943 - 2014)에게 감사드리며, 특히 영어 부분에서 논문 작성 중에 자료와 영어원서에서 나오는 표현을 부드럽게 번역을 해주고 중국어 발음으로 이 교재의 작성에 조언을 해주고 도와준 조은혜에게도 고마움을 전한다.

끝으로 많은 편집시간으로 지체된 작업을 포기하지 않고 완성할 수 있게 해 주신 지오북스의 김남우 사장님과 편집부 직원 여러분의 노고에 진심으로 감사를 드린다.

2022년 9월
조민형

CONTENTS

❖ 머리말 / 1

Chapter 1	영어용어를 한글용어로	5
Chapter 2	영어교과서에 나타나는 수학용어	107
Chapter 3	수의 명칭	189
Chapter 4	도량형 변환과 근삿값	213
Chapter 5	수학 수식 읽기(수식과 기호 읽기)	225
Chapter 6	영어에서 나타나는 수학용어	236
Chapter 7	제2외국어 수학 학습 자료들	251
Chapter 8	알아두면 유용한 것들	299
	참고문헌	313

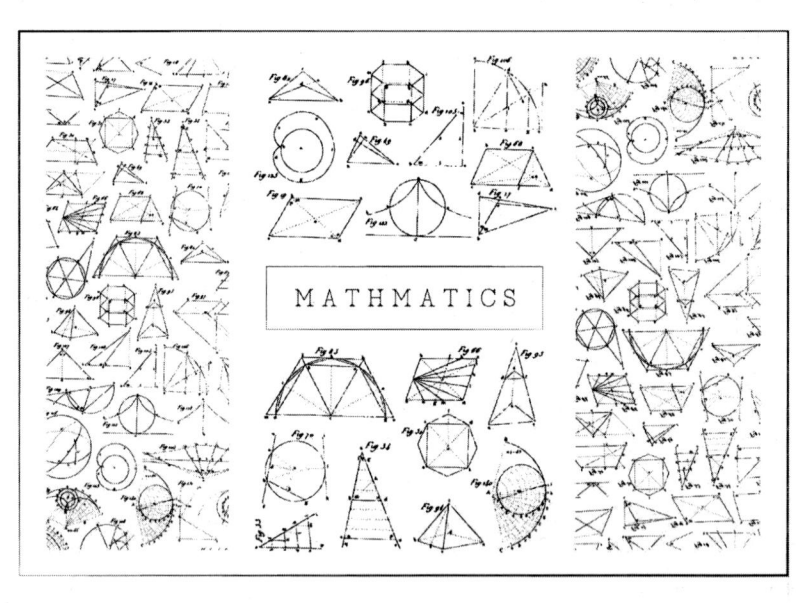

Chapter 1 영어용어를 한글용어로

Chapter 1 영어용어를 한글용어로

이 장에서는 주로 초등수학에서 대학 기초 미분적분학까지 수학에 대한 용어를 모아서 정리하였다. 그러나 나라마다 조금씩 다르겠지만 미적분을 학습하기 전에 삼각함수와 기초 확률·통계에 대한 수업이 이루어지는 경우가 많이 있으며, 이런 것을 고려하여 삼각함수와 확률·통계에 대한 용어가 추가 되었으며, 집합과 논리를 위한 용어와 고등수학에서 다루는 용어의 일부가 부분적으로 추가되었다. 용어 비교를 위해 ↔ 는 서로 반대되는 용어를, (cf) 는 서로 비교되는 용어를 나타내는 것으로 사용하였다.

원서읽기를 위한 수학용어사전

abacus	주판(珠板)
abnormal series	비정규급수(非正規級數)
abscissa	가로좌표, 횡좌표(橫座標), x좌표 ↔ 세로좌표(ordinate)
absolute convergence	절대수렴(絕對收斂)
absolute error	절대오차(絕對誤差) (cf) relative error
absolute inequality	절대부등식(絕對不等式)
absolute value	절댓값
absolutely convergent series	절대수렴급수(絕代收斂級數)
abstract algebra	추상대수학(抽象代數學)
absurdity	불합리(不合理), 모순(矛盾)
abundant number	과잉수(過剩數), 초과수, 잉여수
acceleration	가속도(加速度)
accumulation point	집적점(集積點)
accuracy	정확도(正確度)
acute	예각(銳角)의
acute (angled) triangle	예각삼각형(銳角三角形), acute triangle
acute angle	예각(銳角) (cf) obtuse angle (둔각)
add	더하기, 더하다, 보태다
addend	가수(加數)
addition	덧셈, 더하기
addition or subtraction method	가감(加減)법

Chapter 1 영어용어를 한글용어로

addition property for inequality	부등식(不等式)의 덧셈법칙
addition property of equality	등식(等式)의 덧셈법칙
addition theorem	덧셈정리, 가법정리(加法定理)
additive formula	덧셈공식, 가법공식, 더하기공식
additive function	가법함수(加法函數), 더하기함수
additive inverse	덧셈에서의 역수(逆數)
adjacent	인접(隣接)한 (이웃하는)
adjacent angle	인접각(隣接角)
adjacent arcs	인접호(隣接弧)
adjoin	인접하다, 결합하다
adjoint	딸림, 수반(隨伴), 동반(同伴)
adjoint matrix	수반행렬(隨伴行列), 딸림행렬
algebra	대수(代數), 대수학(代數學)
algebraic	대수적(代數的) (cf) transcendental
algebraic curve	대수곡선(代數曲線)
algebraic expression	대수 표기법, 대수식(代數式)
algebraic number	대수적수 (cf) transcendental number
algebraic operating system	대수연산체계(代數演算體系)
algorithm	셈법, 알고리즘
almost all	거의 모든
almost everywhere	거의 모든 점에서

alternate (interior) angles	엇각(錯角)
alternate exterior angles	외엇각
alternate interior angles	내엇각
alternating expression	교대식(交代式)
alternating series	교대급수(交代級數)
altitude	높이(=height)
altitude of a cone	원뿔의 높이
altitude of a cylinder	원기둥의 높이
altitude of a parallelogram	평행사변형(平行四邊形)의 높이
altitude of a prism	각기둥의 높이
altitude of a pyramid	각뿔의 높이
altitude of a triangle	삼각형의 높이
amicable pair	친화쌍(親和雙), 친화짝
amplitude	진폭(振幅)
analysis	해석(解析), 해석학(解析學), 분석(分析)
analysis of variance	분산분석(分散分析), 분산분석법
analytic function	해석함수(解析函數)
analytic geometry	해석기하학(解析幾何學)
angle	각(角)
angle bisector of a triangle	삼각형의 각의 이등분선(二等分線)
angle of circumference	원둘레각, 원주각, inscribed angle

Chapter 1 영어용어를 한글용어로

angle of intersection	교각(交角)
angular measure	각도(角度)
angular velocity	각속도(角速度)
annulus	환형, 원환(圓環), 고리 (복수형 annuli)
antecedent	전항(前項), 앞항 ↔ 후항(consequent)
antiderivative	역도함수, 원시함수(原始函數)
antidifferentiation	적분(積分)
anti-lexicographic order	반사전순서(反辭典順序)
antisymmetry	반대칭(反對對稱)
aphelion	원일점(遠日點) ↔ perihelion
apothem	변심(變心)거리
applied mathematics	응용수학(應用數學) ↔ pure mathematics
apply	적용하다, 응용하다
approximate	가깝게 하다, 어림하다, 약, 대강(大綱)
approximate expression	근사식(近似式)
approximate value	근삿값
approximation	근사(近似), 어림셈
approximation method	근사방법, 근사법
Arabic numeral	아라비아 숫자
arbitrariness	임의(任意)
arbitrary	임의의, 부정(不定)의

원서읽기를 위한 수학용어사전

arbitrary constant	임의의 상수
arc	호(弧) ↔ 현(弦, chord) (cf) major arc, minor arc
arc length	호의 길이
arc measure	호의 값
arc of a chord	현의 호
Archimedean property	아르키메데스 성질(性質)
area	면적(面積), 넓이
area of base	밑면의 넓이
argument	편각(偏角), 변수(變數), 논의, 논증
arithmetic	산수(算數), 산술(算術), 셈
arithmetic mean	산술평균(算術平均)
arithmetic progression	등차수열(等差數列)
arithmetic sequence	등차수열(等差數列)
arithmetic series	등차급수(等差級數)
arrange	정돈하다, 정리하다
arrangement	배열(配列)
array	열, 배열
ascending order of power	오름차순
ascending power	오름차순(次順) ↔ descending power
associative law	결합법칙(結合法則)
associative property	결합법칙(結合法則)

Chapter 1 영어용어를 한글용어로

assumption	가정(假定), 가설(假說) ↔ conclusion
astroid	별모양, 성망형 　(cf) asteroid 소행성(小行星)
asymmetric	불균형[부조화]의, 비대칭의
asymmetric relation	비대칭적 관계(非對稱的 關係)
asymptote	점근선(漸近線), asymptotic line
asymptotic curve	점근곡선(漸近曲線)
asymptotic line	점근선(漸近線), asymptote
augend	피가산수(被加算數)　(cf) addend 가수
augmentation	첨가(添加), 덧붙임
augmented matrix	첨가행렬, 붙임행렬
autonomous	자율적(自律的)
autonomous equation	자율방정식(自律方程式)
autonomous system	자율계(自律系)
auxiliary circle	보조원(補助圓)
auxiliary equation	보조방정식(補助方程式)
auxiliary line	보조선(補助線)
average	평균(平均), mean
average rate of change	평균변화율(平均變化率)
average value	평균값
average velocity	평균속도(平均速度)
axes	좌표축(座標軸), axis의 복수형

axial symmetry	축대칭(軸對稱)
axiom	원리(原理), 공리(公理)
axiom of choice	선택공리(選擇公理)
axiom of completeness	완비공리(完備公理)
axiom of continuity	연속공리(連續公理)
axiom of extension	확장공리(擴張公理)
axiom of extent	확장공리(擴張公理)
axiom of infinity	무한공리(無限公理)
axiom of order	순서공리(順序公理)
axiom of ordinality	순서수(의)공리
axiom of pairing	짝의 공리
axiom of parallels	평행선공리(平行線公理)
axiom of power	멱공리(冪公理)
axiomatic set theory	공리적집합론(公理的集合論) ↔ naive set theory
axis	축, x-axis, y-axis, (axes의 단수형)
axis of abscissas	가로축, 횡축(橫軸)
axis of coordinates	좌표축(座標軸)
axis of ordinates	세로축, 종축(縱軸)
axis of rotation	회전축(回轉軸)
axis of symmetry	대칭축(對稱軸)
back substitution	역대입법(逆代入法)

Chapter 1 영어용어를 한글용어로

balance	균형(均衡)
bar graph	막대그래프
barycenter	중심(重心), 무게 중심
base	밑, 기수(基數), 기선, 밑변(底邊), (로그의) 밑
base angle	밑각(底角)
base angles of an isosceles triangle	이등변삼각형의 밑각
base of an isosceles triangle	이등변삼각형의 밑변(base)
basis	기저(基底), 밑 (bases의 단수형)
benchmark	수준점, (일반적인)기준, 척도, 표준가격
Bernoulli's trials	독립시행(獨立試行)
Berry's paradox	베리의 역설(逆說)
best-fit line	최적선(最適線)
between (a and b)	(a 와 b를 포함한) 사이
between and inclusive	같거나 크다, 같거나 작다
biased sample	편중표본(偏重標本)
biconditional	쌍조건문(雙條件文)
bijective function	전단사함수(全單射函數)
bilateral	두 면이 있는, 좌우 동형의
billion	10억(十億)
binary notation	이진기수법(二進基數法)
binary scale	2진법

원서읽기를 위한 수학용어사전

binary operation	2항연산
binomial	2항식
binomial distribution	2항분포(二項分布)
binomial theorem	2항정리(二項定理)
bijection	전단사(全單射)
bijective function	전단사함수(全單射函數)
bisect	2등분하다, 이분하다
bisector	(선분·각 등의) 2등분선
body of revolution	회전체(回轉體)
Boolean algebra	부울 대수(代數)
bound	유계(有界)시키다
boundary	경계(境界)
bounded	유계(有界)의
bounded above	위로 유계(有界)
bounded below	아래로 유계(有界)
bounded sequence	유계수열(有界數列)
box-and-whisker graph(plot)	상자-수염 그래프(그림)
braces	{ }, 중괄호(中括弧)
brackets	[ ], 대괄호(大括弧)
broken line graph	꺾은선 그래프
bundle	속(束), 다발

Chapter 1 영어용어를 한글용어로

calculate	계산하다
calculation	계산(計算), 셈법, 셈
calculator	계산기(計算機)
calculus	미적분(微積分), 미분학+적분학
calculus of variations	변분법(變分法)
cancellation	소거(消去), 약분(約分)
cancellation law	소거법칙(消去法則)
canonical	표준(적), 표준이 되는
canonical basis	표준기저(標準基底)
canonical form	표준(標準)형, 표준꼴
Cantor diagonal process	칸토르 대각선(對角線) 방법
Cantor ternary function	칸토르 삼진함수(三進函數)
cardinal number	기수(基數) ↔ ordinal number 서수
cardioid	심장형(心臟形)
Cartesian	데카르트(Descartes)의
Cartesian coordinates	데카르트 좌표(座標)
Cartesian product	데카르트 곱
catenary	현수선(懸殊線)
cathetus	직각삼각형에서 직각을 낀 두변, leg
Cauchy sequence	코시 열, 코시 수열(數列)
cell	세포, 칸, 포체(胞體)

원서읽기를 위한 수학용어사전

Celsius temperature scale	섭씨온도눈금 (cf) Fahrenheit
center	중심(中心)
center angle, central angle	중심각(中心角)
center of a circle	중심점
center of a regular polygon	정다각형의 중점
center of a sphere	구(球)의 중점
center of curvature	곡률중심(曲率中心)
center of gravity	무게중심(重心), center of mass
center of mass	무게중심(重心), center of gravity
center of rotation	회전(回傳)의 중심
center of similarity	닮음의 중심
center of symmetry	대칭(對稱)의 중심
centimeter	cm, 센티미터
central angle	중심각(中心角)
central limit theorem	중심극한정리(中心極限整理)
central line	중심선(中心線)
centroid	무게중심
century	세기(世紀), 100년
chain rule	연쇄법칙(連鎖法則)
chance	우연(偶然), 운
chances	(종종 복수형) 가망, 승산, 가능성(可能性)

Chapter 1 영어용어를 한글용어로

change	거스름돈, 변환
change of basis	기저변환(基底變換), 기저 바꿈
change of variables	변수변환(變數變換), 변수 바꿈
characteristic	표수, 특성곡선, (로그의) 지표, 특성
characteristic equation	고유방정식(固有方程式), 특성방정식
characteristic function	특성함수(特性函數)
characteristic polynomial	고유다항식, 특성다항식
characteristic value	특성값, 고윳값, eigenvalue
characteristic vector	특성벡터, 고유벡터, eigenvector
chart	표, 차트
Chebyshev's inequality	체비셰프 부등식(不等式)
check	검산하다
chi-square distribution	카이제곱분포(分包)
choice function	선택함수(選擇函數)
chord	현(弦) ↔ arc (호)
circle	원(圓)
circle graph	원그래프
circle of convergence	수렴원(收斂圓)
circle of curvature	곡률원(曲律圓)
circular cone	원뿔(圓錐)
circular cylinder	원기둥

원서읽기를 위한 수학용어사전

circular cylindrical coordinates	원기둥좌표, 원주좌표
circular disk	원판(圓板)
circular helix	원나선(圓螺旋)
circular measure	호도법(弧度法)
circular permutation	원순열(圓順列)
circular truncated cone	원뿔대(圓-臺)
circulating decimal	순환소수(循環小數), recurring decimal
circulation	순환(循環)
circumcenter	외심(外心) (cf) incenter
circumference	원주(圓周), 원둘레
circumscribe	외접(外接)시키다, 외접하다
circumscribed circle	외접원(外接圓) ↔ inscribed circle
circumscribed polygon	외접다각형(外接多角形)
circumscription	외접(外接)
cissoid of Diocles	디오클레스의 질주선
class	계급(階級)
class interval	계급구간(階級區間)
class mark	계급(階級)값, 계급치(階級値)
classification	분류(分類)
classify	분류하다
clockwise	시계방향으로 ↔ counterclockwise

Chapter 1 영어용어를 한글용어로

closed	닫힌
closed contour	닫힌 경로, 폐경로
closed curve	닫힌곡선, 폐곡선(閉曲線)
closed disk	닫힌 원판, 폐원판(閉圓板)
closed figure	닫힌 도형, 폐도형
closed interval	닫힌구간, 폐구간(閉區間)
closed set	닫힌 집합, 폐집합(閉集合)
closed surface	닫힌곡면, 폐곡면(閉曲面)
closure	닫힘, 폐포(閉包)
code	부호, 코드
coding	부호화, 코딩
codomain	공변역(共變域) (cf) domain, range
coefficient	계수(係數)
coefficient of correlation	상관계수(相關係數)
cofactor	여인수(餘因數)
collection	모임, 집합(集合)
collinear	동일 직선상(의), 공선 (cf) coplanar
column	열(列), 행렬식의 열 ↔ row(행)
column matrix	열행렬(列行列)
column vector	열벡터
combination	조합(組合)

● 19

원서읽기를 위한 수학용어사전

common chord	공통현(共通弦)
common denominator	공통분모(共通分母)
common difference	공차(公差) (cf) common ratio
common divisor	공약수(公約數)
common factor	공약수(公約數), 공통인수(共通因數)
common fraction	상분수
common logarithm	상용로그(常用log)
common measure	공약수(公約數), common divisor
common multiple	공배수(公倍數)
common ratio	공비(公比) (cf) common difference
common tangent	공통접선(共通接線)
commutative law	교환법칙(交換法則)
commutative property	교환법칙(交換法則)
companion matrix	동반행렬(同伴行列)
comparable	비교가능한
comparison test	비교판정법(比較判定法)
compass	컴퍼스
compatible	양립(兩立)하는
complement	여집합(餘集合), 여수, 나머지집합, 여원
complementary	보충하는, 상보적인
complementary angle	90도에 대한 여각(餘角) (cf) supplementary angle

Chapter 1 영어용어를 한글용어로

complementary event	여사건(餘事件)
complementary function	여함수(餘函數)
complementary set	여집합(餘集合)
complete	완비(完備)(의), 완성하다
complete enumeration	전수조사(全數調査)
complete lattice	완비격자(完備格子)
complete ordered field	완비순서체(完備順序體)
completeness	완비(完備性)
completing the square	완전제곱
complex (number) plane	복소평면(複素平面)
complex conjugate	복소켤레, 복소공액
complex fraction	번분수(繁分數)
complex function	복소함수(複素函數)
complex number	복소수(複素數)
complex variable	복소변수(複素變數)
component	성분(成分)
composite function	합성함수(合成函數)
composite number	합성수(合成數), 1과 소수가 아닌 수
composition	합성(合成)
compound event	복합사건(複合事件)
compound interest	복리(複利) ↔ 단리(simple interest)

computation	계산, 셈
compute	계산하다
concave	오목, 요(凹) ↔ convex
concave down	아래로 오목, 위로 볼록
concave function	오목함수
concave up	아래로 볼록, 위로 오목
concentric	중심이 같은
concentric circle	동심원(同心圓)
conclusion	결론(結論) ↔ assumption
condition	조건(條件)
condition of congruence	합동조건(合同條件)
condition of similarity	닮음의 조건
conditional	조건(條件)명제, 조건(條件)문
conditional convergence	조건부수렴(條件附收斂)
conditional equation	조건등식(條件等式)
conditional inequality	조건부등식(條件不等式)
conditional probability	조건부확률(條件附確率)
conditional proposition	조건(條件)명제
conditional statement	조건문(條件文)
cone	원뿔, 원추(圓錐)
confidence interval	신뢰구간(信賴區間)

Chapter 1 영어용어를 한글용어로

confidence level	신뢰(信賴)수준
confidence region	신뢰영역(信賴領域)
confocal	공초(共焦), 초점이 같은
conformal mapping	등각사상(等角寫像)
congruence	(기하 또는 정수론에서의) 합동(合同)
congruence class	합동류(合同類)
congruence condition	합동조건(合同條件)
congruent	(기하 또는 정수론에서의) 합동의
congruent figures	합동 도형(모양)
congruent triangles	합동인 삼각형
congruity	합동(合同)
conic (section)	원뿔곡선, 원추곡선(圓錐曲線)
conjecture	추측(推測)
conjugate(s)	켤레, 공액(共軛)
conjugate complex number	켤레복소수, 공액복소수(共軛複素數)
conjunction	논리(論理)곱
connected	연결(連結)(된) ↔ disconnected
consecutive	계속되는, 연속되는
consecutive even integers	연속적인 짝수
consecutive integers	연속된 정수들, 연속적인 정수
consecutive odd integers	연속적인 홀수

원서읽기를 위한 수학용어사전

consequence	결과(結果)
consequent	후항(後項) ↔ antecedent(전항)
consistency	무모순성(無矛盾性), 일관성(一貫性)
consistent	모순없는, 일관된 ↔ inconsistent
consistent equation	해를 갖는 연립방정식
constant	상수(常數), 일정, 일정한
constant function	상수함수(常數函數)
constant of variation	계수(係數)
constant term	상수항(常數項)
constraint	제약(制約)
construct	작도(作圖)하다, 그리다
constructible	작도가능(作圖可能)한
construction	작도(作圖), 그리기
continued fraction	연분수(連分數)
continuity	연속(連續), 연속성(連續性)
continuous	연속(連續)의
continuous function	연속함수(連續函數)
continuum hypothesis	연속체가설(連續體假說)
contour	경로(經路)
contraction	축약(縮約)
contradict	모순(矛盾)되다

Chapter 1 영어용어를 한글용어로

contradiction	모순(矛盾), 모순명제 ↔ tautology 항진명제
contraposition	(논리) 대우(對偶) (cf) converse, reverse
converge	수렴(收斂)하다 ↔ diverge
convergence	수렴(收斂) ↔ divergence
convergence radius	수렴반경(收斂半徑), radius of convergence
convergence region	수렴영역(收斂領域)
convergence test	수렴판정(收斂判定)
convergent	수렴하는 ↔ divergent
convergent series	수렴급수(收斂級數)
converse	(논리) 역(逆) (cf) reverse, contraposition
convex	볼록(한), 철(凸) ↔ concave
convex function	볼록함수(函數)
convex polygon	볼록다각형(多角形)
convex set	볼록집합(集合)
convolution	포갬, 합성곱
coordinate	좌표(座標) x-coordinate, y-coordinate
coordinate axis	좌표축
coordinate grid	좌표 눈금
coordinate plane	좌표평면(座標平面)
coordinate proof	좌표평면을 이용한 증명
coordinate system	좌표(座標)계

원서읽기를 위한 수학용어사전

coordinate transformation	좌표변환(座標變換)
coordinates	좌표(座標)
coordinates axis	좌표축(座標軸)
coordinates of a point	점의 좌표(座標)
coplanar	동일평면상의 (cf) collinear
coprime	서로 소(素) (cf) relatively prime
corner	모서리, 능(稜)
corollary	따름정리, 계(系)
correlation	상관(相關), 상관관계(相關關係)
correlation coefficient	상관계수(相關係數)
correspondence	대응(對應), 짝
corresponding angle(s)	동위각(同位角)
corresponding side(s)	대응변(對應邊)
cosecant	cosec, 코시컨트(cosec), 여할(餘割)
cosine	cos, 코사인, 여현(餘弦)
cosine rule	코사인법칙(cos 法則)
cost	원가(原價), 가격(價格), 비용(費用)
cotangent	cot, 코탄젠트, 여접(餘接)
count	(수를) 세다. 계산하다
countability axiom	가산공리(可算公理)
countable	셀 수 있는, 가산(可算) ↔ uncountable

Chapter 1 영어용어를 한글용어로

countable set	가산집합(可算集合), 셀 수 있는 집합
counterclockwise, contraclockwise	반시계방향으로 ↔ clockwise
counterexample	반례(反例), 반증(反證)
covariance	공분산(共分散)
covariation	공변동
cresent	활꼴(弓形), segment (of a circle)
criterion	판정기준(判定基準)
critical point	극점(極點), 임계점(臨界點)
cross product	외적(外積), 벡터곱 ↔ inner product
cross ratio	비조화비, 복비(複比)
cross section	절단면(切斷面)
cube	입방체(立方體), 정육면체(正六面體), 세제곱(立方 또는 三乘)
cube root	세제곱근(立方根)
cubed	3승, 세제곱
cubic	삼차(三次)의, 세제곱, 입방(立方)의
cubic centimeter	입방 센티미터
cubic equation	3차방정식(三次方程式)
cubic inch	세제곱 인치
cubic unit	입방단위(立方單位)
cuboid	입방형의, 주사위 모양의
cumulative frequency	누적도수(累積度數)

원서읽기를 위한 수학용어사전

cup	합집합
curvature	곡률(曲律) (cf) torsion (열률)
curve	곡선(曲線)
cusp	뾰족점, 첨점(尖點)
cut	절단(切斷)　(cf) truncation 절단(截斷)
cyclic group	순환군(循環群)
cycloid	사이클로이드, 굴렁쇠선
cylinder	원기둥
cylindrical	원기둥의
cylindrical coordinates	원주좌표(圓柱座標), 원기둥좌표
damping	감쇠(減衰)(의)
data	자료(資料), 데이터, (datum의 복수형)
data fitting	자료 맞추기
data process	자료처리, 데이터처리
De Moivre's theorem	드 무아브르의 정리(定理)
De Morgan's law	드 모르강의 법칙(法則)
decade	10년
decagon	10각형
deci-	데시, $\frac{1}{10}$
decimal	10진수(의), 소수(의)
decimal expansion	십진 전개, 소수 전개

Chapter 1 영어용어를 한글용어로

decimal fraction	소수 분수, 소수(小數)
decimal part	소수부(小數部)
decimal point	소수점(小數點)
decimal system	10진법(十進法)
decrease	감소(減少)
decreasing	감소하는
decreasing function	감소함수(減少函數)
decreasing ratio	감소율(減少率)
decrement	감소(減少), 감소량 ↔ increment
Dedekind cut	데데킨트 절단(切斷)
deduction	연역(演繹), 연역법(演繹法) ↔ induction
deductive reasoning	연역법(演繹法)
deficient number	부족수(不足數)
definite integral	정적분(定積分) ↔ indefinite integral
definition	정의(定義), 뜻
degenerate	퇴화(退化)
degree	도, 차수(次數), 차(次)
degree of a polynomial	다항식의 차수
degree of freedom	자유도(自由度)
degree of scattering	산포도(散布度)
deka-	데카, deca-, 10배

demonstrate	증명하다
demonstration	논증(論證), 증명(證明)
denominator	분모(分母) ↔ numerator 분자
dense	조밀(稠密)한, 촘촘한 ↔ sparse
density	밀도(密度)
denumerable	번호 붙일 수 있는
dependence	종속(從屬)
dependent	종속(從屬)인 ↔ independent
dependent events	종속사건(從屬事件)
dependent variable	종속변수(從屬變數)
depth	깊이
derangement	교란
derivation	미분작용소, 미분(微分), differentiation
derivative	도함수(導函數)
derivative of higher order	고계도함수(高階導函數)
derivative/derived function	도함수(導函數)
derive	~을 끌어내다
derived function	도함수(導函數)
descending central series	내림중심열
descending power	내림차순 ↔ ascending power
describe	기술하다, 말로 설명하다, 그리다(draw)

Chapter 1 영어용어를 한글용어로

determine	(~의 위치를) 정하다
determinant	행렬식(行列式)
develop	전개하다
development figure	전개도(展開圖)
deviation	편차(偏差)
diagonal	대각선의, 대각선(면) (cf) main diagonal　(주대각선)
diagonal element	대각선원소(對角線元素)
diagonal line	대각선(對角線)
diagonal process	대각화 과정
diagonalizable	대각화 가능한
diagonalization	대각화
diagram	도표(圖表), 다이어그램
diameter	지름, 직경(直徑)
dichotomy	이분법(二分法) (cf) bisection
difference	차(差), 나머지, 차집합(差集合) 다름, 차분(差分)
difference equation	차분방정식(差分方程式), 점화식(漸化式)
difference of sets	차집합(差集合)
differences of squares	두 완전제곱수의 차
differentiable	미분가능한
differential	미분(微分)
differential coefficient	미분계수(微分係數)

원서읽기를 위한 수학용어사전

differential equation	미분방정식
differentiate	미분하다
differentiation	미분(微分), derivation, 미분법(微分法)
digit	자릿수, 수자
dilation, dilatation	확장(擴張)
dilute	희석(稀釋)한, 약하게 하다
dimension	차원(次元), 부피, 면적, 크기, 규모
Diophantine equation	디오판투스 방정식
direct proportion	정비례(正比例)
direct variation	정비례 ↔ 반비례(inverse variation)
direction	방향(方向)
direction angle	방향각(方向角)
direction coefficient	방향계수(方向係數)
direction cosine	방향코사인(方向 cos)
direction ratio	방향비(方向比)
directional derivative	방향도함수(方向導函數)
discontinuous	불연속(不連續)
discount	할인(액)
discounting	디스카운트 (할인해 주는)
discover	나타내다, 밝히다
discrete	이산(離散)의, 불연속적인

Chapter 1 영어용어를 한글용어로

discrete distribution	이산분포(離散分布)
discrete random variable	이산확률변수(離散確率變數)
discrete variable	이산변수(離散變數)
discriminant	판별식(判別式)
disjoint	서로 소, 서로 만나지 않는
disjunction	논리합(論理合)
distance	거리
distance between a point and a line	한 점과 선 사이의 거리
distance between two parallel lines	평행한 두 선 사이의 거리
distance formula	두 점 사이의 거리공식
distribution	분포(分布), 초함수
distributive law	분배법칙(分配法則)
distributive property of multiplication	곱셈의 분배법칙
diverge	발산하다 ↔ converge
divergence	발산(發散) ↔ convergence
divergence theorem	발산정리(發散定理)
divergent	발산(發散)하는 ↔ convergent
divide	나누다
divided difference	차분(差分)상
dividend	피제수, 나누어지는 수 (cf) divisor
dividing rational numbers	유리수(有理數)의 나눗셈

원서읽기를 위한 수학용어사전

divisibility	나누어짐
divisible	나누어지는
division	나눗셈, 제법
division property for inequality	부등식의 분배법칙
division property of equality	등식의 분배법칙
divisor	제수(除數), 약수(約數), 나누는 수 (cf) dividend
divisor of zero	영인자(零因子)
dodecagon	12각형(十二角形)
dodecahedron	12면체
domain (of definition)	정의역(定義域)
domination	지배(支配), 우세(優勢)
dot product	스칼라곱, 내적(內積)
dotted line	점선(點線) ↔ 실선(solid line)
double integral	2중적분(二重積分)
dual	쌍대(雙對)
dual space	쌍대공간(雙對空間)
duality	쌍대성(雙對性)
duodecimal	12진법
duplicating the cube	정6면체의 부피를 두 배로 하기
eccentric angle	이심각(異心角)
eccentricity	이심률(異心率)

Chapter 1 영어용어를 한글용어로

echelon form	사다리꼴
edge	변(邊), 모서리, 능(稜)
eigenvalue	고윳값, characteristic value
eigenvector	고유벡터, characteristic vector
element	원소(元素), 원(元)
elementary function	초등함수(初等函數)
elementary operation	기본연산(基本演算)
eliminate	제거하다, 삭제하다
elimination	소거(消去)
elimination method	소거법(消去法)
ellipse	타원(橢圓 또는 楕圓)
ellipsoid	타원면, 타원체(橢圓體)
elliptic	타원의
elliptic function	타원함수(橢圓函數)
empirical probability	경험적 확률
empty event	공사건(空事件)
empty set	공집합(空集合) (cf) vacant set
endpoint	끝점
enlargement	확대(擴大)
ensembles	집합(集合)
entry	성분(成分)

enumerable	셀 수 있는 (cf) countable
envelope	포락선(包絡線), 포락면(包絡面)
epicycloid	바깥굴렁쇠선, 에피사이클로이드 (cf) hypocycloid, cycloid
equal	같다, 같은
equality	등식(等式), 상등(相等)
equally likely outcomes	결과가 나올 확률이 동등함
equation	방정식(方程式) ↔ inequality 부등식
equation in two variables	이원방정식(二元方程式)
equiangular	등각(等角)의
equiangular triangle	정삼각형(正三角形), regular triangle
equiareal transform	등적변형(等積變形)
equidistance	등거리(等距離), 같은 거리
equidistant	같은 거리의(등거리의)
equilateral	등변의
equilateral triangle	등변삼각형, 정삼각형(正三角形)
equipotent	대등(對等), 두 집합사이에 일대일 대응이 존재할 때
equivalence	동치(同値), 동등(同等)
equivalence class	동치류(同値類), 동등류
equivalence relation	동치관계(同値關係)
equivalent	동치인, 동등한
Eratosthenes' sieve	에라토스테네스의 체

Chapter 1 영어용어를 한글용어로

error	오차(誤差), 참값과 근삿값의 차이
error function	오차함수(誤差函數)
escribed circle	방접원(傍接圓)
estimate	어림잡다, 추정값, 평가(하다), 개산(槪算), 추정하다
estimation	추정(推定)
Euclid's postulates	유클리드의 공준(公準)
Euclidean algorithm	유클리드의 호제법(互除法)
Euclidean geometry	유클리드 기하학(幾何學)
Euclidean space	유클리드 공간(空間)
Euler's formula	오일러의 공식(公式)
Euler's function	오일러의 함수(函數)
evaluate	(값을) 구하다, 평가하다
even function	짝함수, 우함수(偶函數)
even number	짝수, 우수(偶數) ↔ odd number
even permutation	짝순열, 짝치환, 우치환(偶置換)
event	사건(事件)
evolute	축폐선(縮閉線) (cf) involute
exact differential	완전미분(完全微分)
exact differential equation	완전미분방정식(完全微分方程式)
example	보기, 예(例), 예제(例題)
excenter	방심(傍心), 방접원(傍接圓)의 중심

existence theorem	존재정리(存在定理)
existential quantifier	존재기호(存在記號), ∃
expansion	전개(展開)
expansion of function	함수의 전개
expectation	기댓값
expectation value	기댓값
expected value	기댓값
experiment	실험(實驗), 시행(施行)
explain	설명하다
explicit function	양함수(陽函數) ↔ implicit function
explore	탐구하다, 조사하다.
exponent	지수(指數), 멱(冪)
exponential	지수의
exponential decay	감가상각(減價償却)
exponential distribution	지수분포(指數分布)
exponential function	지수함수(指數函數)
exponential growth	지수(적)증가
exponents power	지수승
express	(기호·숫자 따위로) 표시하다
expression	식(式), formula
extended real numbers	확장된 실수

Chapter 1 영어용어를 한글용어로

extension	확장(擴張)
exterior	외부, 바깥 ↔ interior
exterior angle	외각(外角)
exterior angles on the same side	동방외각(同傍外角)
exterior common tangent	공통외접선(共通外接線)
external angle	외각(外角)
external division	외분(外分) ↔ internal division 내분(內分)
extraction	추출(抽出)
extraction of the square root	개평, 제곱근풀이
extraneous root	무연근(無緣根)
extrapolation	보외법(補外法), 외삽법(外揷法)
extremal	극값의
extremal value/extremum	극값, 극대 또는 극솟값
extremes	(비례, 비 또는 급수의) 외항 ↔ means
extreme point	극점(極點)
extreme value	극값
face	면(面), 표면(表面)
factor	인수(因數), 인자(因子), 약수, 인수분해하다
factorial	계승(階乘), 계승의
factoring	인수분해(因數分解)
factorization	인수분해(因數分解)

factorization in prime factors	소인수분해(素因數分解)
Fahrenheit temperature scale	화씨온도눈금 (cf) Celsius
fair game	공정한 게임
false	거짓
family	모임, 족(族)
family of curves	곡선족(曲線族), 곡선의 모임
family of sets	집합족(集合族), 집합의 모임
feasible	실행가능 한 ↔ infeasible
Fibonacci numbers	피보나치 수
field	체(體)
figure	도형(圖形), 숫자, 모양, 그림
final term	끝항, 말항(末項)
finite	유한(有限)(의)
finite (terminate) decimal	유한소수(有限小數)
finite difference	유한차분(有限差分)
finite group	유한군(有限群)
finite ordinal	유한서수(有限序數)
finite sequence	유한수열(有限數列)
finite set	유한집합(有限集合)
first axiom of countability	제1가산(可算)공리
first countable axiom	제1가산공리

Chapter 1 영어용어를 한글용어로

first derivative test	일계 도함수판정법(一階導函數判定法)
first term	초항(初項), 첫째항
first-order differential equation	일계 미분방정식(一階 微分方程式)
five centroids of triangles	삼각형의 오심(五心)
fixed point	고정점(固定点)
floating point	부동소수점(浮動小數点)
floating-point representation	부동소수표현
flow chart	순서도(順序圖)
fluid	유체, 유동적인 (cf) solid, liquid
focus	초점(焦點) (cf) 복수형 foci
foot	ft, 피트 (측정단위)
foot of perpendicular	수선(垂線)의 발
force	힘
formula	공식(公式), 식(式), expression
formulas for half angles	반각(半角)의 공식(公式)
formulas of multiplication	곱셈공식(公式)
formulate	공식으로 나타내다, 공식화하다, (문장을) 수식화하다
forward difference	전진차분(前進差分)
forward elimination	전진소거(前進消去)
forward substitution	전진대입(前進代入)
four fundamental rules of arithmetic	사칙(四則), 사칙계산

Fourier analysis	푸리에 해석(解析)
Fourier expansion	푸리에 전개(展開)
Fourier integral	푸리에 적분(積分)
Fourier series	푸리에 급수(級數)
Fourier transformation	푸리에 변환(變換)
fractal	차원 분열(次元分裂) 도형(圖形), 프랙털
fraction	분수(分數)
fractional equation	분수식(分數式), 분수방정식(分數方程式)
fractional expression	분수식(分數式)
fractionalization	작은 부분으로 나눔
frequency	빈도(頻度), 도수, 진동수(振動數), 주파수(周波數)
frequency distribution	도수분포(度數分布)
frequency table	빈도표, 도수분포표(度數分布表)
frustum	원뿔[각뿔]대, 절두체(截頭體)
frustum of pyramid / prismoid	각뿔대
function	함수(函數), 기능
functional analysis	함수해석학(函數解釋學)
functional notation	함수 표기
fundamental sequence	기본열
fundamental solution	기본해(基本解)
fundamental theorem of algebra	대수학의 기본정리

Chapter 1 영어용어를 한글용어로

fundamental theorem of arithmetic	산술의 기본정리
fundamental theorem of integral calculus	미적분학의 기본정리
Galois group	갈루아 군(群)
gambler's ruin	도박꾼의 파산(破散)
game	게임, 놀이
gamma function	감마함수
Gauss divergence theorem	가우스의 발산정리(發散定理)
Gauss quadrature	가우스 구적법(求積法)
Gaussian integer	가우스 정수
GCD(greatest common divisor)	최대공약수(最大公約數)
general solution	일반해(一般解)
general term	일반항(一般項)
generalized continuum hypothesis	일반 연속체가설(連續體假說)
generating function	생성함수(生成函數), 모함수(母函數)
generating line	모선(母線)
generator	모선(母線)
geoboard	지오보드
geometric(al) mean	기하평균(幾何平均), 등비중항(等比中項)
geometric probability	기하확률(幾何確率)
geometric progression	등비수열(等比數列)
geometric sequence	등비수열(等比數列)

geometric series	등비급수(等比級數), 기하급수(幾何級數)
geometry	기하(幾何), 기하학(幾何學)
global	전체적인 ↔ local
golden section	황금분할(黃金分割)
gradient	기울기, 그래디언트, 물매
gradient vector	기울기벡터
gram	g, 그램
graph	그래프, 그래프를 그리다
graph theory	그래프 이론(理論)
graphic method	그래프를 사용해서 푸는 방법(식)
graphing calculators	그래픽 계산기
great circle	반구의 면
greater than	>, 보다 더 큰
greater than or equal, no less than	≥, 보다 더 크거나 같다
greatest common divisor	GCD, 최대공약수(最大公約數)
greatest element	최대원소(最大元素)
greatest lower bound	glb, 하한(下限), 최대하계(最大下界)
Green's theorem	그린의 정리
gross profit	매상총수익, 순매상고에서 매상원가를 뺀 값
group	군(群), 군론(群論)
group representation	군의 표현(表現)

Chapter 1 영어용어를 한글용어로

half angle	반각(半角)
half line	반직선(半直線)
half-plane	반평면(半平面)
harmonic conjugate	조화켤레, 조화공액(調和共軛)
harmonic function	조화함수(調和函數)
harmonic mean	조화평균(調和平均)
harmonic progression	조화수열(調和數列)
harmonic sequence	조화수열(調和數列)
harmonic series	조화급수(調和級數)
Hausdorff Maximal Principle	하우스도르프 극대원칙(極大原則)
height	높이 = altitude, 키
helix	나선(螺旋)
hemisphere	반구(半球)
heptagon	7각형(七角形)
Hesse's normal form	헤세의 표준형(標準形)
Hessian matrix	헤시안 행렬
hexadecimal	16진법
hexagon	6각형(六角形), 6변형
hexahedron	육면체(六面體)
histogram	히스토그램, 기둥그림표
homeomorphic	위상동형(位相同型) (의)

homeomorphism	위상동형사상(位相同型寫像)
homogeneous	동차의, 동질의 (cf) nonhomogeneous
homogeneous equation	동차방정식(同次方程式)
homogeneous function	동차함수(同次函數)
homomorphism	준동형사상(準同型寫像)
horizontal	가로의, 수평(水平)인
horizontal axis	수평축, x축
hundreds	백의 자리
hundredths	소수 둘째 자리
hyperbola	쌍곡선(雙曲線 또는 双曲線)
hyperbolic	쌍곡선의
hyperbolic function	쌍곡선함수(雙曲線函數)
hypergeometric distribution	초기하분포(超幾何分布)
hypergeometric series	초기하 급수(超幾何級數)
hyperplane	초평면(超平面)
hypotenuse	(직각삼각형의) 빗변(斜邊)
hypothesis	가설(假說), 가정(假定)
icosahedron	이십면체(二十面體)
ideal	아이디얼
idempotent	멱등원(冪等元), 제곱이 같은(원소)
identify	알아내다, 지적하다

Chapter 1 영어용어를 한글용어로

identity	항등식(恒等式), 항등원, 단위원소
identity element	항등원소, 단위원소(單位元素)
identity function	항등함수(恒等函數)
identity matrix	단위행렬(單位行列)
if and only if	필요충분조건(必要充分條件), iff, 동치
if-then statement	조건문(條件文)
ill-conditioned	불량조건의 ↔ well-conditioned
illustrated	그림으로 나타난, 그림으로 설명된
image	상(像)
imaginary axis	허수축(虛數軸)
imaginary number	허수(虛數)
imaginary part	허수부분(虛數部分)
imaginary root	허근(虛根)
imaginary unit	허수단위(虛數單位)
implicit function	음함수(陰函數) ↔ explicit function
implicit solution	음적인 해
imply	내포(內包)하다, 암시(暗示)하다
improper fraction	가분수(假分數)
improper integral	이상적분(異常積分), 특이적분(特異積分)
impulse function	충격함수(衝擊函數)
in ascending power	오름차순(으로)

원서읽기를 위한 수학용어사전

in descending power	내림차순(으로)
incenter	내심(內心) (cf) circumcenter
inch	인치, 2.54cm
incidence matrix	결합행렬(結合行列)
incident angle	입사각(入射角)
inclusion function	포함함수(包含函數)
incompleteness theorem	불완전성 정리(不完全性 定理)
inconsistent system	모순체계, 근을 갖지 않는 연립방정식
increasing	증가의
increasing function	증가함수(增加函數)
increasing ratio	증가율(增加率)
increasing sequence	증가수열(增加數列)
increment	증가분, 증가량, 증분(增分)
indefinite integral	부정적분(不定積分) ↔ definite integral
independence	독립(獨立), 독립성(獨立性)
independent	독립인 ↔ dependent
independent event	독립사건(獨立事件)
independent variable	독립변수(獨立變數)
indeterminate	불확실한, 불확정한
indeterminate form	부정형(不定形)
index	지표(指標), 첨수(添數)

Chapter 1 영어용어를 한글용어로

index number	지수(指數)
indicial equation	결정방정식(決定方程式), 지표방정식
indirect proof	간접증명법(間接證明法)
indirectly proportional	반비례(反比例)
induction	귀납, 귀납법(歸納法) ↔ deduction
induction axiom	귀납공리(歸納公理)
inductive definition	귀납적 정의(歸納的 定義)
inductive reasoning	귀납법(歸納法)
inductive set	귀납집합(歸納集合)
inequality	부등식(不等式) (cf) equation 방정식
inequality of higher degree	고차부등식(高次不等式)
inequality symbol	부등호(不等號)
inertia	관성(慣性)
inferior angle	열각(劣角)
infimum	하한(下限), 최대하계(最大下界)
infinite	무한(의), 무한대(無限大)
infinite decimal	무한소수(無限小數)
infinite geometrical series	무한등비급수(無限等比級數)
infinite recurring decimal	무한순환소수(無限循環小數)
infinite sequence	무한수열(無限數列)
infinite series	무한급수(無限級數)

infinite set	무한집합(無限集合)
infinitesimal	무한소(無限小)
infinity	무한(無限), 무한대(無限大)
inflection point	변곡점(變曲點), inflexion point
initial condition	초기조건(初期條件)
initial line	시초선(始初線)
initial segment	절편(截片)
initial value	초기값
injection	단사(單射), 일대일 사상
injective function	단사함수(單射函數), 일대일 함수
inner center	내심(內心)
inner opposite angle	내대각(內對角)
inner product	내적, 스칼라곱
inscribe	내접시키다, 접하다
inscribed angle	원주각 ↔ central angle(중심각)
inscribed circle	내접원(內接圓)
inscription	내접(內接)
insert	대입하다
integer	정수(整數)
integer part	정수부분
integrable	적분가능한

Chapter 1 영어용어를 한글용어로

integral	정수(整數)의, 적분(積分), 적분의
integral calculus	적분법(積分法)
integral part	정수부분(整數部分)
integral test	적분판정법(積分判定法)
integrand	피적분함수(被積分函數)
integrate	적분하다
integrating factor	적분인자(積分因子)
integration	적분(積分)
integration by parts	부분적분(部分積分), 부분적분법(部分積分法)
integration by substitution	치환적분(置換積分)
integration constant	적분상수(積分常數)
interaction	상호작용(相互作用), 교호작용(交互作用)
intercept	절편(截片), x절편, y절편, x-intercept, y-intercept
interest	이익(利益), 이자(利子)
interior	내부(內部), 안의, 내부의
interior angle	대내각, 내각(內角), 안각
interior angle sum theorem	내각의 합 정리
interior angles on the same side	동측내각(同側內角), 같은 쪽 안각
interior opposite angle	내대각(內對角)
interior point	내부점(內部點), 내점, 안점
intermediate value theorem (IVT)	중간값정리

intermediate variable	중간변수(中間變數)
internal angle	내각(內角)
internal common tangent	공통내접선(共通內接線)
internal division	내분(內分) ↔ external division 외분(外分)
interpolate	보간하다
interpolating polynomial	보간다항식(補間多項式)
interpolation	보간법(補間法)
interquartile range	사분위범위, IQR
intersection (of sets)	교집합(交集合), 교점(交點), 공통집합(共通集合)
intersection angle	교각(交角)
intersection point	교점(交點)
interval	간격(間隔), 구간(區間)
invariant	불변식, 불변량, 불변(不變)
inverse	역(逆), 역수(逆數), 역원(逆元)
inverse element	역원(逆元)
inverse function	역함수(逆函數)
inverse image	역상(逆像)
inverse matrix	역행렬(逆行列)
inverse number	역수(逆數)
inverse operations	역연산(逆演算)
inverse proportion	반비례(反比例)

Chapter 1 영어용어를 한글용어로

inverse trigonometric function	역삼각함수(逆三角函數)
invertible matrix	가역행렬(可逆行列)
investigate	조사하다, 연구하다, 심사하다(examine)
involute	신개선(伸開線) (cf) evolute
involution	대합(對合)
involving	포함하는
irrational equation	무리방정식(無理方程式)
irrational expression	무리식(無理式)
irrational function	무리함수(無理函數)
irrational number	무리수(無理數)
irreducible	기약(의), 나눌 수 없는
irreducible fraction	기약분수(旣約分數)
is approximately equal to	≈, 거의 같은, 근사적으로
isogonal	등각(等角)
isolated	고립(孤立)된
isolated essential singularity	고립된 본질적 특이점, 고립된 진성 특이점
isolated point	고립점(孤立點)
isolated singular point	고립된 특이점
isometry	등거리사상, 등거리변환
isomorphic	동형(同形)
isomorphism	동형(同形)사상

원서읽기를 위한 수학용어사전

isoperimetric problem	등주문제(等周問題)
isosceles	이등변(二等邊)의
isosceles right triangle	직각이등변삼각형(直角二等邊三角形)
isosceles trapezoid	등변(等邊)사다리꼴
isosceles triangle	이등변삼각형(二等邊三角形)
isosceles Triangle Theorem	이등변삼각형 정리
iteration	반복(反復), 되풀이
iteration method	반복법(反復法)
Jacobian	야코비안, 야코비 행렬식
joint distribution	결합분포(結合分布)
joint probability density function	결합확률밀도함수(結合確率密度函數)
Jordan curve	조르당 곡선(曲線)
justify	정당화하다(vindicate), 정당함을 증명하다
kernel	핵(核)
kilo-	k, 킬로
kilogram	kg, 킬로그램
kiloliter	kl, 킬로리터
kilometer	km, 킬로미터
known quantity	기지수(既知數), 알려진 양
Kronecker's delta	크로네커 델타
Lagrange multiplier	라그랑즈 승수(乘數)

Chapter 1 영어용어를 한글용어로

Lagrange multiplier method	라그랑즈 승수법(乘數法)
Laplace equation	라플라스 방정식(方程式)
Laplace transform	라플라스 변환(變換)
Laplacian	라플라스 작용소(作用素)
larger than	>, 보다 더 큰
lateral area	옆면적
lateral faces	옆면, 측면(側面)
Latin square	라틴사각형
lattice	격자(格子), 속(束)
Laurent series	로랑 급수(級數)
law of cosines	코사인법칙
law of exponent	지수법칙(指數法則)
law of great numbers	큰 수의 법칙(法則), 대수법칙(大數法則)
law of inertia	관성법칙(慣性法則)
law of large numbers	대수의 법칙, 큰 수의 법칙
law of parallelogram	평행사변형(平行四邊形)의 법칙
law of proportional part	비례부분의 법칙(比例部分- 法則)
law of sines	사인법칙
least common denominator	최소공분모
least common multiple	LCM, 최소공배수(最小公倍數)
least likely	최소 가능성

least square approximation	최소제곱근사
least square method	최소제곱법, 최소자승법(最小自乘法)
least upper bound	lub, 최소상계, 상한(上限)
least upper bounded	최소상계(最小上界)
Lebesgue integral	르베그적분
left	왼쪽
left inverse	좌역원(左逆元), 왼쪽 역원
left-hand side	LHS, 좌변(左邊), 왼쪽 변 ↔ RHS
leg	직각삼각형에서 빗변을 제외한 두변, 이등변삼각형에서 길이가 같은 두변, 사다리꼴의 평행하지 않은 나머지 두변
Legendre's polynomial	르장드르 다항식
Legendre's symbol	르장드르의 기호
lemma	보조정리(補助定理)
length	길이, 세로 (cf) width
less likely	가능성이 더 적은
less than	<, 보다 적은
less than or equa to, no more than	≤, 보다 더 적거나 같은
level curve	등위선(等位線), 등고선(等高線)
lexicographic	사전편집(辭典編輯)의
lexicographic(al) order	사전순서(辭典順序)

Chapter 1 영어용어를 한글용어로

like denominators	공분모
like terms	동류항(同類項), similar terms
limaçon	리마송, 달팽이꼴곡선
limit	극한(極限), 한계(限界), 극한(값)
limit comparison test	극한비교판정법(極限比較判定法)
limit point	집적점(集積點), 극한점
line	선(線) (cf) straight line 직선
line graph	선 그래프
line integral	선적분(線積分)
line of intersection	교선(交線)
line of symmetry	대칭선 (cf) point of symmetry
line plot	선 도면
line segment	선분(線分)
line symmetry	선대칭(線對稱)
linear	직선의, 1차의, 선형(線型)
linear approximation	일차근사, 선형근사(線型近似)
linear combination	일차결합(一次結合)
linear congruence	일차합동(一次合同)
linear dependence	일차종속(一次從屬)
linear equation	1차 방정식(一次方程式), 선형방정식
linear expression	일차식(一次式)

linear fractional transformation	선형(일차)분수변환
linear function	일차함수(一次函數)
linear graphs	일차함수의 그래프
linear independence	일차독립, 선형독립(線形獨立)
linear operator	선형연산자, 선형작용소(線型作用素)
linear programming	선형계획법(線型計劃法), LP
linear regression	선형회귀(線型回歸), 직선회귀(直線回歸)
linear space	선형공간(線型空間), 벡터공간
linear transformation	선형변환(線型變換), 일차변환
linearly dependent	일차종속인
linearly independent	일차독립인
liquid	액체 (cf) solid, fluid
local	국소적(局所的) ↔ global
local maximum	극대(極大)
local minimum	극소(極小)
locally convex space	국소(局所) 볼록공간
locus	궤적(軌跡), 자취
logarithm	로그, 대수(對數) (cf) 대수(代數)
logarithmic equation	로그방정식(log 方程式)
logarithmic function	대수함수, 로그함수
logarithmic inequality	로그부등식(log 不等式)

Chapter 1 영어용어를 한글용어로

logical symbol	논리기호(論理記號)
loop	닫힌곡선, 고리, 자폐선(自閉線)
lower bound	하계(下界)
lower quartile	제1사분위
Maclaurin series	매클로린 급수(級數)
magic square	마방진(魔方陣)
major axis	장축(長軸), 긴 축
manifold	다양체(多樣體)
manipulation	조작(造作)
mantissa	(로그의)가수(假數) (cf) characteristic(지표)
map, mapping	사상(寫像)
marginal cost	한계비용(限界費用)
marginal density function	주변밀도함수(周邊密度函數)
marginal probability function	주변확률함수
marked down	인하(引下)
marked up	인상(引上)
Markov chain	마르코프 연쇄(連鎖)
Markov process	마르코프 과정(過程)
martingale	마팅게일
mass	무게 重量, 질량(質量), weight
massage	(숫자, 자료를) 조작하다, 분석하다, 마사지(하다)

원서읽기를 위한 수학용어사전

mathematical expectation	수학적 기댓값
mathematical induction	수학적 귀납법(數學的 歸納法)
mathematical probability	수학적 가능성(數學的 可能性)
mathematical probability	수학적 확률(數學的 確率)
mathematical structure	수학적 구조(數學的 構造)
mathematics	수학(數學)
matrix	행렬(行列) (복수형은 matrices)
matrix diagonalization	행렬의 대각화
matrix element	행렬원소(行列元素)
maxima	maximum 의 복수
maximal element	극대원소(極大元素)
maximum	최대(단수), 최댓값 (복수형은 maxima)
maximum point	최대점
maximum principle	최대원리
maximun (point of function)	이차함수의 최대
mean	평균(平均), 평균값, average
means	내항(內項) ↔ 외항(extremes)
mean value theorem	MVT, 평균값정리
measure theory	측도론(測度論)
measurement	측정(測定)(값)
median	중선, 중앙값, 중간값

Chapter 1 영어용어를 한글용어로

median line	중선(中線)
median value	중간값
mensuration by parts	구분구적법(區分求積法)
mental math	암산(暗算)
meter	m, 미터
method of elimination by adding and subtracting	가감법(加減法)
method of equivalence	등치법(等置法)
method of least squares	최소제곱법
method of steepest descent	최급강하법(最急降下法)
method of substitution	대입법(代入法)
method of successive substitution	축차대입법(逐次代入法)
method of undetermined coefficients	미정계수법(未定係數法)
metric units	미터 단위
middle point	중점(中點)
midpoint	(보통 단수취급) 중심점
midpoint (of line segment)	중점
midpoint formulas	중점공식
minima	minimum 의 복수
minimal element	극소원소(極小元素)
minimal polynomial	극소다항식(極小多項式)
minimum	최소, 최솟값 (복수형은 minima)

minimum point	최소점
minimum polynomial	기약다항식(旣約多項式)
minor	소행렬식(小行列式)
minor axis	단축(短軸), 짧은 축
minor determinant	소행렬식(小行列式)
minuend	피감수(被減數)
minus	-, 빼기, 마이너스, 뺄셈
mixed decimal	대소수(帶小數)
mixed fraction	대분수(帶分數)
Möbius strip	뫼비우스의 띠
mode	최빈값, 모드, 최빈수(最頻數)
model	모형(模型), 모델
modulo	법(法)
modulus	(정수론에서의) 법(法)
modus ponens argument	긍정식 논법(肯定式論法)
modus tollens argument	부정식 논법(否定式論法)
moment	모멘트, 적률(積率)
moment generating function	모멘트생성함수, 적률모함수(積率母函數)
monic polynomial	최고차계수가 1인 다항식
monomial	단항식(單項式)
monomorphism	1대1 대응이 있는 homomorphism

Chapter 1 영어용어를 한글용어로

monotone	단조(單調)
monotone convergence theorem	단조수렴정리(單調收斂定理)
monotone decreasing	단조감소(單調減少)
monotone function	단조함수(單調函數)
monotone increasing	단조증가(單調增加)
monotone sequence	단조수열(單調數列)
morphism	사상(寫像)
motion	합동변환, 운동
multiadditive	다중가법의, 다중덧셈의
multiplication	곱셈, 승법(乘法)
multinomial distribution	다항분포(多項分布)
multiple	배수(倍數), 중, 중복(重複)
multiple angle	배각
multiple integral	중적분(重積分)
multiple root	중근(重根)
multiplicand	곱해질 수, 피승수(被乘數)
multiplication	곱셈, 곱하기
multiplication property of equality	곱셈 등식
multiplication table	곱셈표
multiplicity	중복도(重複度)
multiplier	승수(乘數), 곱하는 수

원서읽기를 위한 수학용어사전

multiply	곱셈하기, 곱하다
multiply connected	다중연결(多重連結)
multistep problem	다단계(多段階) 문제
multi-valued function	다가함수(多價函數)
mutually disjoint	둘씩 서로 만나지 않는, 서로소
mutually exclusive	서로 배반적인
n factorial, n!	n의 계승(階乘)
natural logarithm	자연로그(自然 log)
natural number	자연수(自然數), counting number
necessary and sufficient condition	필요충분조건(必要充分條件)
necessary condition	필요조건(必要條件)
negation	부정(否定) (cf) indefinite 부정(不定)
negative	마이너스의, 음(陰)의
negative correlation	음(陰)의 상관관계(相關關係)
negative exponents	음(陰)의 지수(指數)
negative integer	음의 정수
negative number	음수(陰數)
negative semidefinite	음의 준정부호(準正符號)
negative sign	-, 음수부호
neighborhood	근방(近傍)
net	그물, 망(網) (cf) a net price 정가(正價)

Chapter 1 영어용어를 한글용어로

network	회로망(回路網), 네트워크
nilpotent	거듭제곱이 영인
nilpotent element	거듭제곱이 영인 원소
node	마디점, 결절점(結節點), 노드
nonagon	9각형(九角形)
non-commutative	비가환의, 비교환의
non-Euclidean geometry	비유클리드 기하학
nonhomogeneous	비동차의 (cf) homogeneous (동차의)
nonhomogeneous differential equation	비동차 미분방정식
nonlinear	비선형(非線型)
nonnegative definite	음이 아닌 정부호
nonorientable	방향을 줄 수 없는
nonsingular	정칙(正則) ↔ singular(특이)
nonsingular matrix	정칙행렬(正則行列)
nonterminating decimal	무한소수(無限小數)
nontrivial solution	자명하지 않은 해, 비자명해(非自明解)
nonzero	0이 아닌
norm	노름, 노음, 놈
normal	법선(法線), 표준(標準), 정규(正規)
normal distribution	정규분포(正規分布)
normal distribution curve	정규분포곡선(正規分布曲線)

normal equation	표준방정식(標準方程式), 정규방정식
normal line	법선(法線)
normal plane	법평면(法平面)
normal vector	법선벡터
normed space	노음공간
not equal to	≠, ~와 같지 않다
not greater than	~이하(以下)
not less than	~이상(以上)
not smaller than	~이상
notation	표기법(表記法), 기호(記號)
nowhere differentiable function	어디에서도 미분불가능한 함수
n-polygon(convex)	n각형(볼록)
null event	공사건(空事件)
null function	영함수(零函數)
null hypothesis	귀무가설(歸無假說)
null set	공집합(空集合)
null space	영공간(零空間)
nullity	영공간의 차원, 핵공간의 차원
number	수(數), 헤아리다, 번호를 매기다
number line	수직선(數直線)
number theory	정수론(整數論), 수론(數論)

Chapter 1 영어용어를 한글용어로

numeral	숫자
numerator	분자(分子) (cf) denominator 분모
numerical	수의, 숫자상의, 수치(의)
numerical coefficient	수계수(數係數)
numerical expression	수식(數式), 수치적 표현
numerical method	수치계산법(數値計算法)
numerical solution	수치해(數値解)
oblique cone	빗원뿔 (cf) 직원뿔(right cone)
oblique cylinder	빗원기둥
oblique line	사선, 빗금(斜線)
oblique prism	빗각기둥
observation	측정(測定), 관측값
obtuse	둔각의 (cf) acute (예각의)
obtuse angle	둔각(鈍角)
obtuse triangle	둔각삼각형(鈍角三角形)
octagon	8각형(八角形)
octahedral group	8면체군(八面體群)
octahedron	8면체(八面體)
octic group	정사각형 대칭군
odd function	홀함수, 기함수(奇函數)
odd numbers	홀수, 기수(奇數)

odd permutation	홀치환, 기치환(奇置換)
odds	(종종 단수 취급) 차이, 승세, 승산
ogive	누적도수곡선, 누적도수분포도
one fourth	4분의 1
one half	2분의 1
one sided hypothesis	단측가설(單側假說), 한쪽 가설
one third	3분의 1
one-to-one	일대일(一對一)
one-to-one correspondence	일대일대응(一對一對應)
one-to-one function	일대일함수(一對一函數)
one-to-one map	일대일(一對一) 사상
one-to-one mapping	일대일(一對一) 사상
onto	위로의
onto map	위로의 사상, 전사사상(全寫寫像)
open interval	열린구간, 개구간(開區間)
open set	열린 집합, 개집합(開集合)
operation	계산(計算), 연산(演算), 사칙연산(四則演算)
operator	연산자(演算子), 작용소(作用素)
opposite angle	대각(對角), 맞각
opposite edge	대변(對邊), 맞변
opposite side	대변(對邊)

Chapter 1 영어용어를 한글용어로

opposite vertical angles	맞꼭지각
opposites	덧셈에 대한 역원
optimal solution	최적해(最適解)
optimization	최적화(最適化)
or	(집합,논리에서) 또는, (수식에서) 즉
order	위수, 차수(次數), 순서(順序), 계, 차
order isomorphic	순서동형(順序同型)
order of operations	연산(演算)의 순서
order property	순서 성질
order relation	순서관계(順序關係)
ordered field	순서체(順序體)
ordered pairs	순서쌍(順序雙)
ordered set	순서집합(順序集合)
ordering (relation)	순서(관계)
ordinal number	순서수(順序數), 서수(序數), 서수의 ↔ cardinal number (기수)
ordinary differential equation	상미분방정식(常微分方程式), ODE
ordinary point	보통점(普通點)
ordinate	y좌표, 세로좌표 ↔ 가로좌표(abscissa)
organize data	자료(資料)를 정리하다
orientable	방향(方向)을 줄 수 있는
oriented	방향(方向)이 있는, 유향(有向)의

origin	원점(原點), O (O로 표시)
orthocenter	수심(垂心)
orthogonal	직교(直交)
orthogonal function	직교함수(直交函數)
orthogonal projection	정사영(正射影)
orthogonal(ity)	직교(直交)
oscillation	진폭(振幅), 진동(振動)
osculating circle	접촉원(接觸圓)
osculating plane	접촉평면(接觸平面)
outcome	결과(結果), 성과
outer angle	바깥각, 외각(外角)
outer product	외적(外積)
overestimate	과대(過大) 추정값
pair	쌍(雙), 짝
parabola	포물선(拋物線)
parabolic	포물선의
paraboloid	포물면(拋物面)
paradox	역설(逆說), 패러독스
parallel	평행(平行)의, 평행선, 평행한
parallel lines	평행선(平行線)
parallel postulate	평행(平行)정리

Chapter 1 영어용어를 한글용어로

parallel translation	평행이동(平行移動)
parallelepiped	평행6면체(平行六面體)
parallelism	평행(平行)
parallelogram	평행4변형(平行四邊形)
parallelogram law	평행4변형의 법칙
parallelopiped	평행6면체(平行六面體)
parameter	매개변수(媒介變數), 모수(母數), 조변수(助變數)
parameter equation	매개방정식(媒介方程式)
parametric equation	매개(변수)방정식
parent graph	기본 그래프
parenthesis	(), 괄호(括弧), 묶음표 (복수형) parentheses
Parseval equality	파세발의 등식
partial derivative	편도함수(偏導函數)
partial differentiable	편미분(偏微分)
partial differential equation	편미분방정식(偏微分方程式), PDE
partial fraction	부분분수(部分分數)
partially ordered set	순서집합(順序集合)
particular solution	특수해(特殊解) (cf) general solution (일반해)
partition	분할(分割)
Pascal's triangle	파스칼의 삼각형(三角形)
path	경로(經路), 길

Peano Axoims	페아노 공리계
pedal triangle	수선발삼각형
Pell's equation	펠 방정식
penny	1센트
pentagon	5각형, 5변형, 오각형(五角形)
percent	퍼센트(%), 백분율(百分率)
percent of decrease	감소 퍼센트
percent of increase	증가 퍼센트
percent proportion	퍼센트 비율
percentage	백분율(百分率), 퍼센트, 퍼센티지
percentage error	백분율오차(百分率誤差)
perfect number	완전수(完全數)
perfect square	완전제곱
perimeter	주변, 주위, 둘레, 주변 길이, (다각형의)둘레의 길이
period	주기(周期)
periodic function	주기함수(週期函數)
periphery	(원 따위의)둘레
permille	천분율(天分率), 천(千)마다, 천에 대하여
permutation	순열(順列), 치환(置換)
permutation group	치환군(置換群)
perpendicular	직교(直交), 수선(垂線), 수(직)선, 수직면

Chapter 1 영어용어를 한글용어로

perpendicular at midpoint	수직이등분선(垂直二等分線)
perpendicular bisector	수직 이등분선
perpendicular lines	수직선(垂直線)
perpendicular segment	수직인 선분
perpendicular, perpendicularity	수직(垂直)
perspective	원근법(遠近法)
phase	위상(位相), 상(相)
phase plane	위상평면(位相平面)
pi	π, 원주율(圓周率), 원주÷지름
piecewise continuous	조각적으로 연속인, 구간연속인
piecewise smooth curve	조각마다 매끄러운 곡선
pigeonhole principle	비둘기집 원리
pivot	선회축(旋回軸), 추축(樞軸), 피벗
pivotal element	중심원소(中心元素)
plane	평면(平面)
planar	평면(상)의, 2차원의, 평면그래프
plane figure	평면도형(平面圖形)
plane geometry	평면기하학(平面幾何學)
Platonic solid	플라톤 입체(立體), 정다면체(正多面體)
plot	구상하다, 좌표로 위치를 결정하다
plus	＋, 플러스 부호

point	소수점(小數點), 점
point at infinity	무한원점(無限原點)
point of inflection	변곡점(變曲點)
point of reflection	점대칭(點對稱)
point of rotation	회전중심점(回轉中心點)
point of symmetry	대칭점(對稱點) (cf) line of symmetry
point of tangency	접점(接點), tangent point
point symmetry	점대칭(點對稱)
Poisson distribution	푸아송분포
polar	극선(極線), 극선의
polar axis	기선(基線), 극축(極軸)
polar coordinate	극좌표(極座標) (cf) rectangular coordinate
polar form	극형식(極形式)
pole	극(極)
polygon	다각형(多角形), 다변형
polyhedron	다면체(多面體)
polynomial	다항식의, 다항식(多項式)
polynomial distribution	다항분포(多項分布)
polynomial equation	다항방정식(多項方程式)
population	모집단(母集團)
position vector	위치(位置)벡터

Chapter 1 영어용어를 한글용어로

positive	양(陽)의, 양수(陽數)의
positive correlation	양의 상관관계(相關關係)
positive definite form	양의 정부호(正符號)형식
positive direction	양의 방향(方向)
positive integer	양의 정수
positive number	양수(陽數)
positive semidefinite	양의 준정부호
possible outcome	가능한 결과
posteriori	사후, 경험적 ↔ priori
posteriori probability	사후확률(事後確率)
postulate	공리, 공준(公準)
power	멱(冪), 누승(累乘), 거듭제곱
power series	거듭제곱 급수, 멱급수(冪級數)
power series expansion	거듭제곱 급수전개(級數展開)
power set	멱집합(冪集合)
pre-calculus	미적을 배우기 위한 내용으로 구성된 수학
predict	예언하다(prophesy), 예측하다
prediction	예측(豫測)
predictor-corrector method	예측자 수정자법(豫測子 修正子法)
preimage	역상(inverse image) (cf) image 상(像)
premise	전제(前提)

presentation	표시(表示), 표현(表現)
price	대가, 가격
prime	소수(素數), prime number
prime divisor	소인수(素因數), prime factor
prime factor	소인수(素因數)
prime factorization	소인수분해(素因數分解)
primitive	원시(의)
primitive function	원시함수(原始函數)
primitive root	원시근(原始根)
principal	원금(元金), 자본(資本)
principal part	주요 부분, 으뜸 부분
principal value	주치, 주요값
principle	원리(原理)
principle of duality	쌍대원리(雙對原理)
priori	사전, 선험적 ↔ posteriori
priori probability	사전확률, 선험적 확률
prism	각기둥, 삼각기둥, 프리즘
prismoid	각뿔대
probability	확률(確率), 가능성(可能性)
probability density	확률밀도(確率密度)
probability density function	확률밀도함수(確率密度函數)

Chapter 1 영어용어를 한글용어로

probability distribution	확률분포(確率分布)
probability generating function	확률생성함수(確率生成函數)
probability mass function	확률질량함수(確率質量函數)
problem-solving strategies	응용문제 푸는 법
process	과정, 공정, 확률과정(確率過程)
product	곱, 적(積)
product event	곱사건(事件)
product of powers	거듭제곱수의 곱
product property of square roots	제곱근의 곱셈법칙
product set	곱집합
progression	수열(數列)
progression of differences	계차수열(階差數列)
projection	사영(射影)
projective geometry	사영기하학(射影幾何學)
proof	증명(證明)
proper	적절한
proper fraction	진분수(眞分數)
proper subset	진부분집합(眞部分集合)
proportion	비례(比例)
property	성질(性質), 법칙(法則)
proposition	명제(命題), statement

protractor	각도기(角度器)
prove	증명(證明)하다
public opinion survey	여론조사(輿論調査)
pure imaginary number	순허수(純虛數)
pyramid	각뿔, 각추(角錐), 피라미드
Pythagorean number	피타고라스의 수
Pythagorean theorem	피타고라스의 정리(定理)
Pythagorean triple	피타고라스의 수
quadrangle	사각형(四角形)
quadrangular numbers	사각수(四角數)
quadrant	4분원, 4분면(四分面)
quadrate	정방형(正方形)의, 정사각형의
quadratic	2차방정식(二次方程式) ; 2차의
quadratic approximation	이차근사(二次近似)
quadratic convergence	이차수렴(二次收斂)
quadratic equation	이차방정식(二次方程式)
quadratic form	이차형식(二次形式)
quadratic formula	이차방정식의 근의 공식(公式)
quadratic function	2차함수(二次函數)
quadratic residue	제곱잉여(剩餘)
quadratic term	이차항(二次項)

Chapter 1 영어용어를 한글용어로

quadrature	구적법(求積法)
quadrilateral	(평행한 변을 가지고 있지 않은) 4변형, 4각형(四角形)
quantifier	한정기호(限定記號), (전칭기호 + 존재기호)
quantity	양(量)
quart	쿼트(액체의 단위), 1/4 갤론
quartile	4분위수(數), 4분위(分位)의
quasi-	유사(類似), 의사(擬似), 준(準)
quasi-norm	준노름
quaternary	사원의, 사원수의
quaternion	사원수(四元數)
quaternion group	사원수군(四元數群)
quotient of powers	거듭제곱수의 나눗셈
question	질문(質問)
questionnaire	질문지(質問紙)
quinary	5진법(五進法)
quotient	몫, 지수, 상(商)
quotient group	상군(商群), 몫군
quotient map	몫사상
quotient set	몫집합
quotient space	몫공간
radian	라디안, 호도(弧度)

radical equation	무리방정식(無理方程式)
radical expressions	무리식(無理式)
radical root	거듭제곱근
radical sign	근호(根號), 루트($\sqrt{\ }$)
radicand	제곱근 안에 들어 있는 수 또는 수식
radius	반지름, 반경(半徑) (복수형은 radii)
radius of convergence	수렴반경(收斂半徑), 수렴반지름
radius vector	동경(動徑)
raise	제곱하다, 누승하다
random	확률적(確率的), 임의의, 무작위(無作爲)의
random number	난수(亂數)
random sampling	임의추출법(任意抽出法)
random variable	확률변수(確率變數)
range	치역(値域), 범위(範圍)
rank	(행렬의)계수(階數), 순위(順位)
rate	속도(速度), 비율(比率), 율(率)
rate of change	변화율(變化率)
rate of increase	증가율(增加率)
ratio	비(比), 율(率), 비율(比率)
ratio of similarity	닮음비
ratio of the circumference of a circle to its diameter	원주율(圓周率), π

Chapter 1 영어용어를 한글용어로

ratio test	비판정법(比判定法)
rational equations	분수방정식(分數方程式), 유리방정식
rational expression	유리식(有理式), 분수식(分數式)
rational function	유리함수(有理函數)
rational number	유리수(有理數)
rationalization	유리화(有理化)
ray	반직선, 사선(射線)
real axis	실수축(實數軸)
real line	실직선(實直線)
real number	실수(實數)(유리수와 무리수를 합한 수)
real part	실부(實部), 실수부분
real root	실근(實根)
real variable	실변수(實變數)
rearrangement	재배열(再排列, 再配列)
rearrangement theorem	재배열정리(再排列定理, 再配列定理)
reciprocal	곱셈에 대한 역원, 상반된, 역[반대]의, 역수, 상반
reciprocal equation	상반방정식(相反方程式)
reciprocal proportion	반비례(反比例)
rectangle	직사각형(直四角形), 장방형
rectangle graph	직사각형 그래프
rectangular coordinate plane	직교 좌표평면(座標平面)

rectangular coordinates	직교좌표(直交座標) (cf) polar coordinate
rectangular parallelepiped	직육면체(直六面體)
rectangular solid	직육면체
rectifiable	길이를 갖는
rectifiable curve	길이를 갖는 곡선
rectifying plane	전직평면(展直平面)
recurrence formula	점화식(漸化式)
recurrence relation	점화관계(漸化關係)
recurring decimal	순환소수(循環小數), repeating decimal
recursive definition	귀납적 정의(歸納的 定義)
reduce	약분(約分)하다, 통분(通分)하다, 풀다
reduced row-echelon form	기약(既約)행사다리꼴
reductio ad absurdum	귀류법(歸謬法), reduction to absurdity
reduction of a fraction	약분(約分)
reduction to common denominator	통분(通分)
reentrant	안을 향해 있는 ↔ salient 돌출한
refer	참조(參照)하다
refinement	세분(細分), 잘게 나눔
reflection	반사(反射), 뒤집기
reflexive	반사적(反射的)
reflexive law	반사율(反射律)

Chapter 1 영어용어를 한글용어로

reflexive relation	반사관계(反射關係)
region	영역(領域)
region of convergence	수렴영역(收斂領域)
regression	회귀(回歸)
regression analysis	회귀분석(回歸分析)
regression equation	회귀방정식(回歸方程式)
regular	정칙(正則), 정(正)
regular dodecahedron	정12면체(正十二面體)
regular falsi method	오차조정(법), 가위치법(假位置法)
regular function	정칙함수(正則函數)
regular hexagon	정6각형
regular hexahedron	정6면체(正六面體), cube
regular icosahedron	정20면체(正二十面體)
regular octahedron	정8면체(正八面體)
regular pentagon	정5각형
regular polygon	정다각형(正多角形)
regular polyhedron	정다면체(正多面體), Platonic Solid
regular prism	정각기둥
regular pyramid	정각뿔
regular singularity	정칙특이점
regular tetrahedron	정4면체

regular triangle	정3각형(正三角形), equiangular triangle
rejection region	기각영역(棄却領域)
relation	관계(關係)
relative error	상대오차(相對誤差) (cf) absolute error
relative extrema	극값
relative frequency	상대도수(相對度數)
relative maximum	극댓값
relative minimum	극솟값
relatively prime	서로 소(素), 서로 나눌 수 없는 (cf) coprime
reliability	신뢰성(信賴性), 신뢰도(信賴度)
remainder	나머지, 잉여(剩餘), 잔차(殘差)
remainder theorem	나머지정리
remote interior angles	주어진 외각과 접해 있지 않은 두 내각
removable singular point	없앨 수 있는 특이점, 제거가능 특이점
repeating decimal	순환소수, recurring decimal
replicated sampling	반복추출(反復抽出)
represent	말하다, 기술하다, 표현하다
representation	표현(表現)
representative value	대푯값
residual	잉여의, 잉여(剩餘), 잔차
residue	유수(留數), 나머지, 잉여(剩餘)

Chapter 1 영어용어를 한글용어로

residue theorem	유수 정리(留數 定理)
restriction	제한(制限)
reverse	(논리) 이(裏) (cf) converse, contraposition
revolution	회전(回轉), 주기(週期)
revolve	회전(回轉)하다
rhombus	마름모, 사방형(斜方形)
Riemann integral	리만 적분(積分)
right	직각의 (cf) oblique 비스듬한
right angle	직각(直角)
right angled triangle	직각삼각형(直角三角形), right triangle
right circular cone	직원뿔
right circular cylinder	직원기둥
right cylinder	직각기둥
right hand law	오른손 법칙(法則)
right handed coordinate system	오른손 좌표계
right prism	직각기둥
right pyramid	직각뿔
right triangle	직각삼각형(直角三角形), right angled triangle
right-angled triangle	직각삼각형(直角三角形)
right-hand side	RHS, 우변(右邊), 오른쪽 변 ↔ LHS
ring	환(環)

rise	수직으로의 변화 ↔ run
risk	위험(危險)
Rolle's theorem	롤의 정리(定理)
root	근(根), 답, 해(解)
root test	근판정법(根判定法)
rotate	회전하다
rotation	회전(回轉)
rotation axis	회전축(回轉軸)
round number	어림수
round off	반올림
rounding off	반올림
round-off error	반올림 오차(誤差)
row	행(行) ↔ column (열)
row vector	행벡터
row-echelon form	행 사다리꼴
rule of thumb	주먹구구식 계산, 대충, 대략
ruler	자
run	수평으로의 변화 ↔ rise
Russell's paradox	러셀의 역설(逆說)
saddle point	안장(鞍裝)점
salient	돌출한 ↔ reentrant 안을 향해 있는

Chapter 1 영어용어를 한글용어로

sample	표본(標本)
sample mean	표본평균(標本平均)
sample space	표본공간(標本空間)
sample survey	표본조사(標本調査)
sample variance	표본분산(標本分散)
sampling	표본조사(標本調査)
sampling with replacement	복원추출(復原抽出)
sampling without replacement	비복원추출(非復原抽出)
sandwich theorem	샌드위치 정리(定理)
scalar	스칼라, 상수(실수)　(cf) vector
scale	눈금, 축척(縮尺)
scalene	(삼각형이) 부등변의, (원뿔의) 축이 비스듬한
scalene triangle	부등변삼각형(不等邊三角形)
Schwarz's inequality	슈바르츠 부등식(不等式)
scientific notation	유효숫자 표기법, 십진법의 표기법
secant	할선(割線), 시컨트(sec), 정할(正割)
secant line	할선(割線)
second	제2의, 초
second (order) derivative	이계도함수(二階導函數)
second derivative test	이계도함수 판정(判定)
section	단면(斷面), 자른 면, 절단(切斷)

section paper, graph paper	모눈종이
sector	부채꼴(扇形), (cf) 사립문 선(扇)
segment	(직선의)선분(線分)
segment of a circle	활꼴(弓形), cresent
self-adjoint	자기수반(自己隨伴)
semicircle	반원, 반원형
separable	분해가능한, 분리가능한
separation axiom	분리공리(分離公理)
separation constant	분리상수(分離常數)
separation of variables	변수분리법(變數分離法)
septagon	7각형(七角形)
sequence	수열(數列)
sequence of point	점렬(點列)
series	급수(級數)
series of positive terms	양항급수(陽項級數)
series solution	급수해(級數解)
set	집합(集合)
set-builder form	조건제시법(條件提示法) ↔ tabular form
set theory	집합론(集合論)
side	변(邊) (cf) leg
sides of the equation	방정식(方程式)의 변

Chapter 1 영어용어를 한글용어로

sieve	체, 조리
sign	부호(符號)
significance level	유의수준(有意水準)
significant digit	유효수자(有效數字)
similar	닮은꼴의, 닮은, 닮음
similar figures	닮은꼴
similar polygons	닮은 다각형(多角形)
similar solids	닮은 입체(立體)
similar terms	동류항(同類項), like terms
similar transformation	닮음변환
similar triangles	닮은 삼각형(三角形)
similarity	닮음
simple closed curve	단일폐곡선(單一閉曲線)
simple curve	단순곡선(單純曲線)
simple interest	단리(單利) ↔ 복리(compound interest)
simplest form of expression	간단히 한 식
simplify	정리하다, 간단히 하다
Simpson's formula	심프슨의 공식
Simpson's rule	심프슨의 법칙
simultaneous differential equation	연립미분방정식(聯立微分方程式)
simultaneous equations	연립방정식(聯立方程式)

원서읽기를 위한 수학용어사전

simultaneous inequalities	연립부등식(聯立不等式)
sine	sin, 사인, 정현(正弦)
sine curve	사인곡선
sine rule	사인법칙(法則)
single valued function	일가함수(一價函數), 한값함수
singleton	한원소 집합
singular matrix	특이행렬(特異行列)
singular point	특이점(特異點)
singularity	특이점(特異點)
situation vector	위치벡터(位置-)
skew lines	비대칭선(非對稱線)
skew position	꼬인위치(位置)
skew symmetric	반대칭, 의대칭, 왜대칭
skew symmetric matrix	왜대칭 행렬, 교대행렬(交代行列)
skewness	일그러짐, 비틀림
slant	비스듬한, 경사진
slice of a solid	단면(斷面)
slide rule	계산자
slope	기울기
smooth curve	매끄러운 곡선(曲線)
solid	입체(立體)(의), 고체(의), 단단한

Chapter 1 영어용어를 한글용어로

solid figure	입체도형(立體圖形)
solid line	실선(實線) ↔ 점선(dotted line)
solid geometry	입체기하학(立體幾何學)
solution	해법(解法), 해(解), 풀이
solution of a system of equations	연립방정식(聯立方程式)의 해
solution root	해(解), 답(答)
solution set	해집합(解集合)
solution space	해공간(解空間)
solve (for x)	(x 에 대해) 풀이하다
solving a triangle	삼각형 풀기
space	공간(空間)
space figure	공간도형(空間圖形)
span	생성(生成), 펼침
spanned	생성된
sparse	희박(稀薄)한, 드문드문한
sparse matrix	희박한 행렬 ↔ dense matrix
specify	상술(詳述)하다, 자세히 쓰다
spectral radius	스펙트럼 반경(半徑)
speed	빠르기, 속력(速力)
sphere	구(球)
spherical angle	구면각(球面角)

spherical coordinate	구면좌표(球面座標)
spiral	소용돌이선, 나선(螺線)
spline	스플라인, 운형(雲形)자
spreadsheets	컴퓨터용 회계처리장부(會計處理帳簿)
square	정방형(正方形), 정사각형 (正四角形), regular quadrilateral, 제곱하다, 제곱(의)
square free integer	제곱 인수가 없는 정수
square matrix	정사각행렬, 정방행렬
square number	정사각수(正四角數)
square pyramid	사각뿔
square root	제곱근(平方根)
squaring the circle	원과 같은 넓이의 정사각형 작도
squeeze theorem	조임정리, 압축정리(壓縮定理)
stability	안정성(安定性)
standard	표준(標準), 기준(基準)
standard deviation	표준편차(標準偏差)
standard form	표준형(標準形)
standard normal distribution	표준(標準)정규분포
star-shaped	별모양(의)
state	상태, 식으로 나타내다, 진술하다
statement	명제(命題), proposition
stationary point	안정점(安定點)

Chapter 1 영어용어를 한글용어로

statistic	통계량(統計量)
statistic data	통계자료
statistical analysis	통계분석(統計分析)
statistical hypothesis	통계적 가설(假說)
statistical inference	통계적 추론(推論)
statistical probability	통계적 확률(確率)
statistics	통계학, 통계(統計)
steady state	안정상태(安定狀態)
stem-and-leaf plot	줄기-잎 도표
step function	계단함수(階段函數)
Stieltjes' integral	스틸체스 적분(積分)
stochastic process	확률(確率)과정
stochastic variable	확률변수(確率變數)
Stokes' theorem	스토크스의 정리
straight angle	평각(平角)
straight line	직선(直線)
straightedge	직선자
structure	구조(構造)
subadditivity	준가법성
subgroup	부분군(部分群)
subordinate	종속(從屬)된

subsequence	부분수열, 부분열
subset	부분집합(部分集合)
subspace	부분공간(部分空間)
substitute	치환(置換)하다
substitution	대입(代入), 치환(置換)
subtend	(현(弦)·변(邊)이 호(弧)·각(角)에) 대하다
subtract	빼다
subtraction	뺄셈
subtraction property for inequality	부등식의 뺄셈법칙
subtraction property of equality	등식의 뺄셈법칙
subtrahend	감수(減數), 빼는 수
successive approximation method	축차근사법(逐次近似法)
sufficient condition	충분조건(充分條件)
sufficiency	충분성(充分性)
sufficient	충분한
sufficient condition	충분조건(充分條件)
sum	합(合)
sum of sets	합집합(合集合)
sum of squares	제곱합
summand	피가수, 가수, 합의 한 항
summation	합

Chapter 1 영어용어를 한글용어로

superposition	겹침, 중첩(重疊)
superset	포함집합(包含集合)
supplementary	보각의, 보조(補助)
supplementary angle	180도에대한보각(補角) (cf) complementary angle
supremum	상한(上限), 최소상계(最小上界)
surface	곡면(曲面)
surface area	표면적(表面的), 겉넓이
surface integral	면적분(面積分)
surjection	전사함수, 전사(全射), 위로의 함수
surjective	전사의
surjective function	전사함수(全射函數)
syllable	음절(音節)
syllogism	삼단논법(三段論法)
symbol	기호(記號)
symbolic logic	기호논리학(記號論理學)
symmetric	대칭(對稱)의, 대칭적(對稱的)
symmetric difference	대칭차(對稱差)
symmetric law	대칭률(對稱律)
symmetric matrix	대칭행렬(對稱行列)
symmetric transposition	대칭이동(對稱移動)
symmetry	대칭(對稱)

synthetic division	조립(組立)제법
system of axiom	공리체계(公理體系)
system of coordinates	좌표(座標)계
systems of equations	연립방정식(聯立方程式)
systems of inequality	연립부등식(聯立不等式)
table	표(表)
table of square roots	제곱근표(平方根表)
tabular form	원소나열법(元素羅列法) ↔ set-builder form
tabulate	표로 작성하다
tally	세다, 기록하다
tangent	tan, 접선(接線), 정접(正接), 탄젠트, 접하는
tangent line	접선(接線)
tangent plane	접평면(接平面)
tangent point	접점(接點), point of tangency
tangent segment	접선(接線)
tangent vector	접선벡터
tangram	칠교판(七巧板)
tautology	항진명제, 항상 참인 명제 ↔ 모순 명제 contradiction
Taylor series	테일러 급수(級數)
temporary average	가평균(假平均)
tens	십(十)의 자리

Chapter 1 영어용어를 한글용어로

tensor	텐서
tenth	소수 첫째자리
term	항(項)
terminating decimal	유한소수(有限小數)
ternary	삼진의, 셋 사이의
tessellation	쪽 맞추기
test of hypothesis	가설검정(假說檢定)
tetrahedron	사면체(四面體)
the area of a circle	원의 넓이
The Axiom of Choice	선택(選擇)공리
The Axiom of Existence	존재(存在)공리
The Axiom of Extensionality	확장(擴張)공리
The Axiom of Infinity	무한(無限)공리
The Axiom of Pair	짝공리
The Axiom of Power Set	멱집합공리
The Axiom of Union	합집합공리
The Axiom Schema of Comprehension	함축(含蓄)공리
The Axiom Schema of Replacement	치환공리(置換公理)
the circumference of a circle	원의 둘레, 원주(圓周)
the intersection of A and B	A와 B의 교집합(交集合)
the point of tangency	접점(接點)

the union of A and B	A와 B의 합집합(合集合)
theorem	정리(定理)
theorem of intermediate value	중간값의 정리
theorem of three perpendiculars	삼수선(三垂線)의 정리(定理)
thousands	천(千)의 자리
thousandth	소수 셋째 자리
three-dimensional	3차원적, 입체적(立體的)
time series	시계열(時系列)
times	곱하기, 몇 배
tolerance	허용(許容)
tolerance interval	허용구간(許容區間)
topological space	위상공간(位相空間)
topology	위상수학(位相數學), 위상(位相)
torsion	비틀림(률), 열률(捩率), 제2곡률
torus	도넛 모양, 원환체(圓環體), 원환면, 토러스
total	합계(合計)
total differential	전미분(全微分)
total event	(행렬의) 전사건(全事件)
trace	(행렬의)대각합(對角合)
tractrix	(행렬의)추적선(追跡線)
trajectory	자취, 궤적(軌跡)

Chapter 1 영어용어를 한글용어로

transcendental function	초월함수(超越函數) (cf) algebraic number
transcendental number	초월수(超越數)
transfinite induction	초한귀납법(超限歸納法)
transfinite ordinal	초유한 서수
transform	변환(變換)
transformation	변환(變換)
transient	일시적인
transitive	추이적(推移的)
transitive law	추이율(推移律)
transitive property of equality	추이성, (a=b이고 b=c이면, a=c이다.)
translation	평행이동(平行移動)
transpose	전치행렬, 전치(轉置), 이항(移項)하다
transpose of a matrix	행렬의 전치(轉置)
transposed matrix	전치행렬(轉置行列)
transposition	호환(互換), 자리바꿈
transposition matrix	호환행렬(互換行列)
transversal	횡단선, 횡단하는
trapezium	(미국) 부등변 사각형(의), (영국) 사다리꼴(의)
trapezoid	(미국) 사다리꼴(의), (영국) 부등변 사각형(의)
trapezoid rule	사다리꼴공식
trapezoidal rule	사다리꼴공식

tree	수형도(樹型圖), 나뭇가지 그림
tree diagram	수형도(樹型圖)
trial	시행(試行)
trial and error	시행착오(試行錯誤)
triangle	삼각형(三角形)
triangle inequality	삼각부등식(三角不等式)
triangular cylinder	삼각기둥
triangular number	삼각수(三角數)
triangular prism	삼각기둥
triangular pyramid	삼각뿔
trichotomy	삼분법(三分法)
trigonometric equation	삼각방정식(三角方程式)
trigonometric function	삼각함수(三角函數)
trigonometric inequality	삼각부등식(三角不等式)
trigonometric ratios	삼각비(三角比)
trigonometric series	삼각급수(三角級數)
trigonometric table	삼각함수표(三角函數表)
trigonometry	삼각법(三角法)
trihedron	삼면체(三面體)
trinomial	3항식, 3항(의)
triple	3배의, 3중의, 세겹의

Chapter 1 영어용어를 한글용어로

triple integral	삼중적분(三重積分)
triple product	삼중적(三重積)
triple scalar product	스칼라 삼중적(三重積)
trisecting the angle	각의 삼등분
trivial	자명한, 사소한
trochoid	트로코이드
true	참, 진(眞)
truncate	절단(切斷)하다
truncated (circular) cone	원뿔대(圓錐臺)
truncated pyramid	각뿔대(角錐臺)
truncation error	절단오차(切斷誤差)
truth set	진리집합(眞理集合)
truth table	진리(眞理)표
truth value	진릿값
twin primes	쌍둥이 소수(素數)
two sided hypothesis	양측가설(兩側假說)
two sided test	양측검정(兩側檢定)
two-dimensional	2차원의, 평면적인
type 1 error	제1종 오류(誤謬)
type 2 error	제2종 오류(誤謬)
unary operation	일항연산(一項演算)

원서읽기를 위한 수학용어사전

unbiased estimate	불편추정값
unbounded set	유계(有界)가 아닌 집합
uncountability	셀 수 없음, 비가산성
uncountable	셀 수 없는, 비가산 ↔ countable
uncountable set	셀 수 없는 집합, 비가산 집합
undecidable	결정불가능(決定不可能)
undefined notion	무정의개념(無定義概念)
undefined term	무정의 용어(用語)
undetermined coefficient	미정계수(未定係數)
uniform	균등(均等)
uniform boundedness	고른 유계성, 균등 유계성
uniform continuity	고른 연속(성), 균등연속(성)
uniform convergence	고른 수렴(성), 균등수렴(성)
uniform distribution	고른 분포, 균등분포(均等分布)
unimodular	유니모듈라
union	합집합(合集合), sum of sets
unique	유일한, 일의적인
unique factorization domain	유일한 인수분해 정역
unique factorization theorem	유일한 소인수 분해정리
unique solution	유일한 해(解)
uniqueness	유일성(唯一性)

Chapter 1 영어용어를 한글용어로

unit	단위(單位)
unit area	단위넓이
unit circle	단위원(單位圓)
unit cost	단위 값, 단가(單價)
unit element	단위원소(單位元素)
unit fraction	단위분수(單位分數)
unit matrix	단위행렬(單位行列)
unitary matrix	유니타리행렬
units	일의 자리
universal	보편적인
universal quantifier	전칭기호(全稱記號), $\forall$
universal set	전체집합(全體集合)
unknown	미지수(未知數) (cf) variable 변수
unknown quantity, unknown variable	미지수, 변수(變數)
upper base	윗변
upper bound	상계(上界)
upper limit	위끝, 상극한
upper quartile	제 3사분위
use	사용, 쓰다
vacant set	공집합(空集合) (cf) empty set
value	값, 수치(數値)

value of function	함숫값
vanish	값이 영(0)이 되다, 값이 없어지다
variable	변수(變數)
variance	분산(分散)
variant	변화, 변형, 일치하지 않는
variation	진폭(振幅), 변동(變動), 변분
vector	벡터
vector analysis	벡터 해석(解析)
vector space	벡터공간(空間)
velocity	속도(速度)
velocity vector	속도벡터
Venn diagram	벤 다이어그램, 벤 도표
verbal expression	대수식을 말 또는 문장으로 적은 것
verify	검증(檢證)하다
versus	~과 대비해서, vs, in contrast with
vertex	꼭짓점, 정점(頂點) (복수형은 vertices)
vertex angle	꼭지각, 정각(頂角)
vertical	수직인
vertical angles	맞꼭지각, 대정각(對頂角)
vertical axis	수직축 ↔ horizontal axis (수평축)
vertical bar graph	수직 막대그래프

Chapter 1 영어용어를 한글용어로

vertically opposite angles	맞꼭지각(角)
volume	체적(體積), 부피
wave equation	파동방정식(波動方程式)
weak	약한
weight	무게, 중량(重量), 질량(質量), mass
weight function	무게함수(函數), 가중치 함수
weighted average	가중평균(加重平均)
weighted mean	가중평균(加重平均)
well defined	잘 정의된
well ordered set	정렬집합(整列集合)
well ordering	정렬순서(整列順序)
well-order	정렬(整列)
whole number	자연수에 0을 포함한 집합, 전수
width	가로, 너비, 폭
winding number	회전수(回轉數)
word square	(세로나 가로로 읽어도 같은 말이 되는) 정사각형의 말의 배열
Wronskian	론스키 행렬식
x-axis	x축
x-coordinate	x좌표(座標)
x-intercept	x절편(切片)
yard	야드(=36인치, 3피트, 약 0.914미터)

원서읽기를 위한 수학용어사전

y-axis	y축
y-coordinate	y좌표(座標)
y-intercept	y절편(切片)
z-axis	z축
zenithal angle	천정각
Zermelo well-ordering theorem	체르멜로 정렬정리
zero	영(零)
zero divisor	영인자(零因子)
zero exponent	0인 지수
zero matrix	영행렬(零行列)
zero product property	어느 수에 0을 곱하면 0이 된다
zero vector	영(零)벡터
zeros (of a function)	함수를 0으로 만드는 값, 근
ZF Axioms	ZF 공리계(公理界)
ZFC Axioms	ZFC 공리계
Zorn's Lemma	초른 도움정리

Chapter 2 영어교과서에 나타나는 수학용어

Chapter 2 영어교과서에 나타나는 수학용어

　이 장에서는 미국 중고등학교에서 배우는 수학과 일부 대학 기초수학에서 나오는 용어를 문장 속에서 정리하였다. 한글은 직역보다는 이해를 돕기 위해 의역을 하였으며 생략하기도 하였다. 대학에서 배우는 수학교재는 쉽게 원서로 접할 수 있기 때문에, 미국 초중등 수학의 범위 안에서 다루고 있으며 쉬운 내용으로 해당하는 부분마다 설명을 위해 복습으로 같은 내용이 중복된다.

　~ing 로 시작하는 분사구문은 접속사가 있는 부사절(접속사, 주어, 동사)을 부사구로 만드는데, 예를 들어 As we simplify the fraction, we have the following equation. (분수를 간단히 하면, 다음 식을 얻을 수 있다.) 라는 문장을 Simplifying the fraction, ~ (분수를 간단히 하면, ~)와 같이 주절을 생략하고 ~ 으로 나타내기로 한다. 따라서 ~ 부분은 스스로 보충하여 문장을 만들어 보면 좋을 것 같다.

2.1 기초수학

● 앞으로 읽으나 뒤로 읽으나 같은 수를 회문의 수 (palindrome number)라고 한다.

: A number that reads the same backward and forward is called a palindrome.

● (반올림, round off) 어떤 자리에서 반올림한다는 것은 그보다 하나 오른쪽 자릿수가 5 이상이면 어떤 자리의 수를 하나 올리고 4 이하이면 그대로 둔다.

: To round to a certain place,
(a) locate the digit in that place;
(b) then consider the digit to the right;
(c) if the digit to the right is 5 or higher, round up;
　　if the digit to the right is less than 5, round down.

● 3872.2459를 소수점 아래 셋째자리, 둘째자리, 첫째자리, 1의 자리, 10의 자리, 100의 자리, 천의 자리에서 반올림한 수를 각각 구하여라.

: Round 3872.2459 to the nearest thousandth, hundredth, tenth, one, ten, hundred, and thousand.

(answer) 3872.246, 3872.25, 3872.2, 3872, 3870, 3900 and 4000

Chapter 2 영어교과서에 나타나는 수학용어

배수를 구하는 방법

☞ A number is divisible by 2 (is even) if it has a one's digit of 0, 2, 4, 6, or 8 (has an even one's digit).

☞ A number is divisible by 9 if the sum of the digits is divisible by 9.

☞ A number is divisible by 3 if the sum of the digits is divisible by 3.

☞ A number is divisible by 4 if the number named by the last two digits is divisible by 4.

☞ A number is divisible by 8 if the number named by the last three digits is divisible by 8.

☞ For a number to be divisible by 6, the sum of the digits must be divisible by 3 and the one's digit must be 0, 2, 4, 6, or 8 (even).

☞ A number is divisible by 10 if the one's digit is 0.

☞ A number is divisible by 5 if the one's digit is 0 or 5.

☞ A number is divisible by 2^n if the number consisting of the last n figures is divisible by 2^n.

- $\dfrac{n}{n} = 1$, for any whole number n that is not 0.

  $\dfrac{0}{n} = 0$, for any whole number n that is not 0.

$\dfrac{n}{1} = n$, for any whole number n.

● 어떤 수에 1을 곱하면 그 수 자신이 된다.
: When we multiply a number by 1, we get the same number.

● 49 ≠ 48 read "49 is not the same as 48."
여기서 read [발음은 과거형 red] 는 읽는다, 읽힌다.

● 곱해서 1 이 되는 두 수를 역(reciprocal)이라고 하고, 어떤 수의 역을 구하려면 분자와 분모를 교환하면 된다.
: If the product of two numbers is 1, we say they are reciprocals of each other. To find a reciprocal, interchange the numerator(분자) and denominator(분모).

● 0은 역수가 없다 ! : 0 has no reciprocal !

● To divide, multiply the dividend(피제수, 나눔수) by the reciprocal of the divisor(제수, 나눗수).

● 분모가 같은 수의 덧셈과 뺄셈은 분모는 그대로 두고 분자끼리 더하거나 뺀다.
: To add [subtract] when denominators are the same,
  (a) add [subtract] the numerator, and
  (b) keep the denominator.

● The LCM of two natural numbers is the smallest

Chapter 2 영어교과서에 나타나는 수학용어

number that is a multiple of both.

● To determine which of two numbers is greater when there is a common denominator, compare the numerators.

● 이미 간 거리 + 앞으로 갈 거리 = 전체 거리
 : Distance already traveled + Distance to go
    = Total distance ran

● 공통분모를 구하여라. : Find a common denominator.
분모는 그대로 둔다. : Keep the denominator.
분수를 대분수로 고치려면 (분자를 분모로) 나눈다.
: To convert from fractional notation to a mixed numeral, divide.

● proper and improper fraction

☞ improper fraction (가분수(假分數)) : 예를 들어 $\frac{8}{5}$

☞ proper fraction (진분수(眞分數)) : 예를 들어 $\frac{2}{5}$

☞ mixed number (대분수(帶分數)) : 예를 들어 $2\frac{1}{5}$

주의 대분수(帶分數)에서 帶(대)는 한자로 큰 대(大)가 아니다. 대(帶)는 '띠다', '허리에 차다' 라는 뜻이다.

● borrow (뺄셈에서 윗자리에서 빌려 오다, 꾸어오다)

(예를 들어 "41 − 12"를 계산할 경우) 마지막 자리 1에서 2를 뺄 수 없으므로 10의 자리에서 하나를 빌려와서 11에서 2를 뺀다.

: But 1 is smaller than 2, so we cannot subtract until we borrow. Borrowing tens to subtract ones, ~

● 그림에서 길이 d를 구하여라.
: Find the length d in this figure.

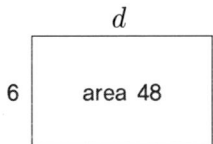

● 분수와 소수가 섞여있는 형태의 수를 곱하거나 나누려면 먼저 분수로 고친다. 나눗셈의 경우는 분자와 분모를 바꿔서 곱하는 것에 주의

: To multiply using mixed numerals, we first write fractional notation.

: To divide using mixed numerals, we first write fractional notation.

: Remember to multiply by the reciprocal.

● 보통 체온은 화씨 98.6도인데 열이 있어서 화씨 107도에 이르면 사망하게 된다.

: Normal body temperature is 98.6°F. When fevered, most

Chapter 2 영어교과서에 나타나는 수학용어

people will die if their bodies reach 107°F.

● A student bought a book for $14.68 and paid with a $20 bill. How much change was there?

● The rectangular page in a book measures 23.2 centimeters by 21.8 centimeters. Find the area.

● 자동차로 4시간 동안에 250마일을 갔다. 한 시간 동안에는 얼마를 갔는가?
 : A car went 250 miles in 4 hr. How far did it go in 1 hr?

● $\frac{5}{6}$ 를 소수로 고쳐라. : Find decimal notation for $\frac{5}{6}$.
 (answer) $0.8\overline{3}$ = point eight three recurring
    (또는 recurring 대신에 repeating)

● Find the quotient(몫) and whole-number remainder(나머지).

● Convert to
  Convert from dollars to cents. : 달러를 센트로 환산하여라.
  Convert to a mixed numeral(대분수). : 대분수로 고쳐라.
  Convert to decimal notation. : 소수로 바꾸어라.

- $\dfrac{3}{2} = \dfrac{6}{4}$ read(읽힌다) '3 is to 2 as 6 is to 4'

- A unit price(단가) is the price of one of something.
  $$\text{unit price} = \dfrac{\text{price}}{\text{number of units}}$$

- (어느 축척지도에서) $\dfrac{1}{4}$ 인치는 실제거리 50 마일을 나타낸다.

  : On a map inch represents 50 actual miles.

- $38\% = \dfrac{38}{100}$: A ratio of 38 to 100

퍼센트는 100 에 대한 어떤 수의 비이다.

: A percent is a ratio of some number to 100 .

- 양변에 역수를 취한다.

: Taking the reciprocal of both sides.

지수 읽기

p^n ⋯ p to the n-th power

5^n ⋯ five to the n-th power

5^{n+1} ⋯ five to the n plus first power

5^3 ⋯ five cubed, the cube of five

5^4 ⋯ five to the fourth power

Chapter 2 영어교과서에 나타나는 수학용어

4^2 … "4 squared" 또는 "4 square"

10^5 … 5 is the exponent(지수), 10 is the base(밑)
$\sqrt{36} = 6$ read(읽힌다) "The square root of 36 is 6"
$28\ km^2$ … 28 square kilometers

☞ We agree that $a = a^1$, for any base a.
☞ We agree that $1 = a^0$, if a is not zero.
  (= Any nonzero number to the 0 power is 1.)
☞ If $a = c^2$, then c is a square root of a.

이 장의 처음에서 언급한 분사구문이 포함된 문장이다.

☞ Simplifying the fraction, ~ : 분수를 간단히 하면 ~
☞ Taking square root, ~ : 제곱근을 취하면 ~
☞ Approximate $\sqrt{27}$ to three decimal places.
: $\sqrt{27}$ 을 소수 셋째자리까지 근사시켜라.

1 yard = 3 feet
1 mile = 5280 feet = 1760 야드 = 약 1,609 미터
1 inch = 2.54 cm

참고 1 마일은 걸어서 대략 20분 걸리는 거리이다.

☞ Substituting 3 ft for 1 yd, ~
: 1야드 대신에 3피트를 대입하면 ~
☞ Substituting 24 hr for 1 day, ~
: 하루를 24시간으로 바꾸면 ~

● It will help to memorize these.
☞ The perimeter(둘레) of a rectangle is twice the length plus twice the width.
$$P = 2(l + w)$$
☞ The perimeter of a square is four times the length of a side.
$$P = 4 \cdot s$$
☞ The area of a rectangular region is the product of the length and the width.
$$A = l \cdot w$$
☞ The area of a square region (정사각형 영역) is the square of a length of a side.
$$A = s^2$$
☞ A parallelogram(평행사변형) is a four-sided figure with two pairs of parallel sides.
☞ The area of a parallelogram is the product of the length of a base and the height.
$$A = b \cdot h$$
☞ The area of a triangle is half the length of the base

Chapter 2 영어교과서에 나타나는 수학용어

times the height.
$$A = \frac{1}{2}b \cdot h$$

☞ A trapezoid(사다리꼴) is a four-sided figure with at least one pair of parallel sides.

☞ The area of a trapezoid is half the product of the height and the sum of the lengths of the parallel sides.
$$A = \frac{1}{2} \cdot h \cdot (a + b)$$

☞ The circumference(원주) of a circle is the distance around it.

☞ The area of a circle with radius(반지름) of length r is given by
$$A = \pi r^2$$

☞ The volume of a rectangular solid is the number of unit cubes it takes to fill it.

☞ The volume of a rectangular solid is found by multiplying length by width by height.
$$V = l \cdot w \cdot h$$

☞ 1 milliliter = $\frac{1}{1000}$ liter

☞ A milliliter is about $\frac{1}{5}$ of a teaspoon(찻숟가락).

American system

1 ton = 2000 pounds (lb)

1 kg ≈ 2.2 lb

1 metric ton = 2200 lb

1 lb = 16 ounces (oz)

long ton = 1,016kg,   short ton = 907kg,

metric ton = 1,000kg

1 square foot (ft^2) = 144 square inches (in^2)

1 square yard (yd^2) = 9 square feet (ft^2)

1 acre = 43,560 ft^2

1 square miles (mi^2) = 640 acres

1 hectare (ha) = 100 are = 100 m × 100 m

1 are (a) = 100 m^2 = 10 m × 10 m

참고 A well-known hamburger is called a "quarter pounder." A "quarter pounder" can also be called a "four ouncer." (쿼터파운더는 1/4 파운드 고기를 사용한 햄버거이다.)

● There is a difference between mass(질량) and weight(무게). Weight is related to the force of gravity(중력). The farther you are from the center of earth, the less you weigh. Your mass stays the same no matter where you are.

Chapter 2 영어교과서에 나타나는 수학용어

● To go from g to Kg in the table is a move of 3 places to the left. Thus, we move the decimal point 3 places to the left.

● A microgram is one-millionth of a gram.
$$\left(1\,\mu g = \frac{1}{1,000,000}\,g\right)$$
One million micrograms is one gram.
$$(1,000,000\,\mu g = 1\,g)$$

● Celsius(섭씨)와 Fahrenheit(화씨)
Convert to Celsius. (섭씨로 바꾸어라.)
Convert to Fahrenheit. (화씨로 바꾸어라.)

Convert from Celsius temperature to Fahrenheit, and from Fahrenheit to Celsius using the formulas

☞ (섭씨를 화씨로) $F = \dfrac{9}{5} \cdot C + 32$

Multiply by $\dfrac{9}{5}$ and add 32. : $\dfrac{9}{5}$를 곱한 다음 32를 더한다.

☞ (화씨를 섭씨로) $C = \dfrac{5}{9}(F - 32)$

Subtract 32 and multiply by $\dfrac{5}{9}$. : 32를 뺀 다음 $\dfrac{5}{9}$를 곱한다.

참고 A. Celsius (1701-1744) 스웨덴의 천문학자
G.D. Fahrenheit (1686-1736) 독일의 물리학자

● 양수, 음수, 절댓값, 역수

☞ Zero is not considered either positive or negative.

☞ The absolute value(절댓값) of an integer can be thought of as its distance from 0 on the number line.

☞ The absolute value of a positive number stays positive.

☞ The absolute value of a negative number is positive.

☞ The absolute value of 0 is 0.

☞ To do the addition $a + b$, we start at a, and move according to b.

   (a) If b is positive, we move to the right.

   (b) If b is negative, we move to the left.

   (c) If b is 0, we do not move.

☞ To add two negative integers, we add their absolute values and change the sign (the answer is negative).

☞ To add a positive and a negative integer, find the difference of their absolute values.

   (a) If the negative integer has the greater absolute value, the answer is negative.

   (b) If the positive integer has the greater absolute value, the answer is positive.

☞ The difference $a - b$ is the number which when added to b gives a.

Chapter 2 영어교과서에 나타나는 수학용어

주의

영어로 수식을 읽을 때 minus 와 negative 는 다음과 같이 구별되어 읽어야 한다.

① $-1-5$: negative one minus five

② $-3-(-7)$: negative three minus negative seven

③ $-a$ 를 negative a 라고 읽어서는 안 된다. 왜냐하면 a 가 negative 이면 $-a$ 는 positive 이기 때문이다.

④ (+2) + (+3) = (+5)

: Positive 2 plus Positive 3 equals Positive 5

⑤ (+6) − (+3) = (+3)

: Positive 6 minus Positive 3 equals Positive 3

⑥ (+6) + (−3) = (+3)

: Positive 6 plus Negative 3 equals Positive 3

⑦ 위 ⑤와 ⑥에 의하여

Subtracting a positive is the same as adding a negative.

⑧ $9-5$ 를 수식으로 불러주면 nine minus five 이다. 이것을 nine negative five 로 불러주면 상대방은 두 개의 수 9과 -5 를 나란히 적게 된다. 즉 $9, -5$ 가 된다.

☞ To multiply two negative integers, multiply their absolute values. The answer is positive.

☞ Any nonzero number a has a reciprocal(역수) $\dfrac{1}{a}$. The reciprocal of $\dfrac{a}{b}$ is $\dfrac{b}{a}$.

☞ The reciprocal of a negative number is negative.
: 음수의 역수는 음수이다.
☞ The reciprocal of a positive number is positive.
: 양수의 역수는 양수이다.

☞ "The distributive law of multiplication over subtraction"
: 뺄셈에 대한 곱셈의 분배법칙
For any rational numbers a, b, and c,
$a(b-c) = ab - ac$

● 인수분해, 지수법칙
☞ Factor out the largest common factor. (인수분해 할 때는 최대공약수를 빼낸다.)
☞ Collect like terms. (동류항을 모은다.)
동류항 = like terms = similar terms
☞ Factor out the x.
☞ If n is any positive integer, b^{-n} is given the meaning $\dfrac{1}{b^n}$. In other words, b^n and b^{-n} are reciprocals(역수).
☞ In multiplication with exponential notation, we can add exponents if the bases are the same :
$a^m \cdot a^n = a^{m+n}$.

Chapter 2 영어교과서에 나타나는 수학용어

☞ In division, we subtract exponents if the bases are the same : $\dfrac{a^m}{a^n} = a^{m-n}$.

Subtracting exponents, ~

: (지수 나눗셈에서는) 지수끼리 빼면, ~

Carry out the subtraction.

☞ The above is true whether the exponents are positive, negative, or zero.

☞ To raise a power to a power we can multiply the exponents. For any exponents m and n,

$(a^m)^n = a^{mn}$.

☞ When several factors are multiplied in parentheses, raise each factor to the given power :

$(a^m a^n)^t = (a^m)^t (a^n)^t = a^{mt} a^{nt}$.

● If principal(원금) P is invested at interest(이자) rate r, compounded(이자를 복리로) annually, in t years it will grow to the amount A given by $A = P(1+r)^t$.

● Scientific notation(과학적 표기) for a number consists of exponential notation for a power of 10 and, if needed, decimal notation for a number between 1 and 10, and a multiplication sign : $N \times 10^n$ or 10^n.

● 암산

mental arithmetic calculation, mental arithmetic computation
☞ Convert mentally to decimal notation.
: 암산을 이용하여 소수로 나타내다.
☞ convert A to B : A를 B로 바꾸다

● The replacements that make an equation true are called solutions(해, 풀이). To solve an equation means to find all its solutions.

● The addition principle : If an equation $a = b$ is true, then $a + c = b + c$ is true for any number c.

● $x - 8.4 = 13.5$ 에서 x 를 구하여라.
: Solve for x, $x - 8.4 = 13.5$.
좌변에 -8.4 를 없애기 위해 양변에 8.4 를 더한다.
: Adding 8.4 to get rid of -8.4 on the left
$x - 8.4 + 8.4 = 13.5 + 8.4 \Rightarrow x = 21.9$

주의 이와 같이 미국 교과서에는 "이항한다"는 말 대신에 양변에 같은 수를 더하거나 빼서 이항하는 것과 같은 효과를 얻을 수 있음을 보인다.

● The multiplication principle : If an equation $a = b$ is

Chapter 2 영어교과서에 나타나는 수학용어

true, then $a \cdot c = b \cdot c$ is true for any number c.
Multiplying by $-\frac{4}{5}$ to get rid of $-\frac{5}{4}$ on the right

● Dividing on both sides by -1 (양변에 -1 로 나누면)
cleared the fractions : 분수를 상수배하여 정수를 만듦
cleared the decimals : 소수를 상수배하여 정수를 만듦

● 양변에 (어떤 수를) 곱하여 분수나 소수를 없앤다. (즉, 정수가 되도록 한다).
: Multiply on both sides to clear of fractions or decimals.

● 필요하다면 양변에 동류항을 모은다. 동류항은 같은 변으로 모은다.
: Collect like terms on each side, if necessary.

● 변수를 가지는 항을 한쪽 변으로 모으고 나머지 항 들을 다른 변으로 모은다.
: Get all terms with variables on one side and all the other terms on the other side.

● Divide to solve for the variable.
Translate (바꾸다) to an equation.
Solve the equation.
Check (점검하다, 검산하다) the answer in the original

problem.

☞ $7 - 3 = 4$

: 7 objects, take away 3 objects, 4 objects remain.

☞ $7 + x = 15$

: 7 plus what number is 15 ?

● borrow : 뺄셈에서 윗자리에서 꾸어오다, 빌려오다
☞ Borrow a ten. : 10의 자리에서 하나(10)를 꾸어온다.
☞ Borrow a hundred.
: 100의 자리에서 하나(100)를 꾸어온다.
☞ Borrowing tens to subtract ones.
☞ Borrowing hundreds to subtract tens.
☞ 9에서 7을 뺄 수 없으므로 10의 자리에서 빌려온다.
: We cannot subtract 9 from 7. We borrow a ten.
: We cannot subtract 60 from 40. We borrow a hundred.

● 어떤 수에 0을 곱하면 0 이 된다.
: Multiplying by 0 gives 0
● (어떤 수에) 1을 곱하면 그 값은 변하지 않는다.
: Multiplying a number by 1 doesn't change it
● Any number divided by 1 is that same number.
● To multiply by 10, write 0 on the end.
: 어떤 수에 10을 곱하면 그 수의 끝자리에 0을 한 개 붙인다.

Chapter 2 영어교과서에 나타나는 수학용어

: To multiply by 100, write 00 on the end.
: To multiply by 1000, write 000 on the end.

> **주의** 896 × 347 은 Multiply 896 by 347 이라하며 "896을 347배" 하는 것이다.

- Division by 0 is not defined.
: 0으로 나누는 것은 정의되지 않는다.
☞ We agree not to divide by 0.
☞ Zero divided by any number greater than zero is zero.
☞ Any number greater than zero divided by itself is 1.

2.2 중급수학

- 물건을 셀 때 사용하는 수를 자연수(natural number)라고 한다.
: The numbers used for counting are called natural numbers.

> **주의** 수학 용어에서 자연수, 자연로그, 자연지수, 자연방정식 등등. 이 들의 공통점은 정말로 일상생활에서 자연스럽게 나타나는 것들에 대해 natural 이란 명칭을 사용하였음에 주목할 필요가 있다.

- 전수(whole number)란 자연수와 0을 합한 수이다.
: The whole numbers consist of the natural numbers and zero.

- 10^3 : ten cubed 또는 ten to the third power

We call the number 3 an exponent (지수) and we say that 10 is the base (밑수).

- 10^2 : ten squared 또는 ten to the second power
- If n is 0, then n^0 is meaningless.

: 0^0 은 무의미하다. 0^0 은 정의하지 않는다.

0.64	decimal notation	(소수 표기)
$\dfrac{16}{25}$	fractional notation	(분수 표기)
0.8^2	exponential notation	(지수 표기)
64%	percent notation	(백분율 표기)

- A letter that can be replaced by different numbers is called a variable (변수).

- 소괄호 ( ), 중괄호 { }, 대괄호 [ ]

각각 영어로 parenthesis, brace, bracket 이라 하는데 중괄호와 대괄호는 경우에 따라 바꿔 불리거나 바꿔 사용되는 경우가 있으며, 괄호가 괄호 안에 나타날 때는 제일 안쪽에 있는 괄호부터 먼저 계산한다.

: When parentheses occur within parentheses, the computations in the inner ones are to be done first.

Chapter 2 영어교과서에 나타나는 수학용어

● 수식을 인수분해하는 것은 그것을 (인수의) 곱으로 나타내는 것이다.

: To factor an expression is to write it as a product.

● 수식에서 더하기 또는 빼기에 의해 나누어진 각 부분을 항(term)이라고 한다.

: In the expression $x + y + z$, the parts separated by plus signs are called terms.

● 단리 : simple interest
　복리 : compound interest

Simple interest (단리) on a principal of P dollars invested at interest rate r for t years is given by Prt.

● principal(원금) + interest(이자) = amount(원리 합계)

● 두 항이 같은 문자를 가지면 그들을 동류항이라고 한다. 우리는 동류항을 모아서 간단히 할 수 있다.

: If two terms have the same letters, they are called like terms (동류항), or similar terms. We can often simplify expressions by collecting or combining like terms.

● 곱해서 1 이 되는 두 수는 서로의 역이라 한다. 0을 제외한 모든 수는 역을 가지고 있다.

: Two numbers whose product is 1 are called reciprocals of each other. All the numbers of arithmetic, except zero, have reciprocals.

- To divide, multiply by the reciprocal of the divisor :
$$\frac{a}{b} \div \frac{c}{d} = \frac{a}{b} \cdot \frac{d}{c}$$

- For any numbers a and b,

    $a < b$ ("a is less than b."로 읽음)

means that a is to the left of b of the number line(수직선).

- For any numbers a and b,

    $a > b$ ("a is greater than b."로 읽음)

means that a is to the right of b of the number line.

- 방정식과 해

☞ An equation (방정식) is a number sentence with = for its verb (동사).

☞ Some equations are true. Some are false. Some are neither true nor false.

☞ The replacements that make an equation true are called solutions.

☞ 방정식을 푼다는 것은 그의 모든 해를 구하는 것을 말한다.

: To solve an equation means to find all of its solutions.

Chapter 2 영어교과서에 나타나는 수학용어

● $x + a = b$를 풀기 위해서는 양변에서 a를 뺀다.
(미국 교과서에서는 "이항한다"는 말을 직접적으로 사용하지 않는다.)

☞ To solve $x + a = b$, subtract a from both sides.

☞ To solve $ax = b$ when a is nonzero, divide on both sides by a (or multiply on both sides by $\frac{1}{a}$).

● We never divide by zero. : 어떤 수를 0 으로 나눌 수 없다.

● $x + 37 = 73$

: What number plus thirty-seven is seventy-three?

● $\frac{3}{4}x = 35$

: Three-fourths of what number is thirty-five?

● The area of Alaska(미국에서 가장 큰 주) is about 483 times the area of Rhode Island(미국에서 가장 작은 주).

● rectangle (직사각형)

☞ If a rectangle has length l and width w, the area is given by

$A = lw$ (Area is length times width.)

☞ The perimeter(둘레의 길이) of a rectangle of length l and width w is given by

$P = 2l + 2w$, or $2(l + w)$

☞ If the length and width of a rectangle are the same, then the rectangle is a square.

☞ In a square, all four sides are the same length.

☞ If a square has sides of length s, then the area is given by $A = s^2$ and the perimeter is given by $P = 4s$.

● parallelogram (평행사변형) and trapezoid (사다리꼴)

☞ A parallelogram is a four-sided figure with two pairs of parallel sides.

☞ The area of a parallelogram is given by $A = b \cdot h$, where b is the length of the base and h is the height.

☞ A trapezoid is a four-sided figure with at least one pair of parallel sides.

☞ If a trapezoid has bases of lengths a and b and has height h, its area is given by

$$A = \frac{1}{2} \cdot h \cdot (a + b).$$

주의 trapezium [trəpíːziəm]은 미국에서는 부등변사각형을 말하고 영국에서는 사다리꼴을 말한다. trapezoid [trǽpəzɔid]는 영국에서는 부등변사각형을 말하고 미국에서는 사다리꼴을 말한다.

● triangle (삼각형)

☞ If a triangle has a base (밑변) of length b and has

Chapter 2 영어교과서에 나타나는 수학용어

height (높이) h, then the area (넓이) is given by $A = \dfrac{1}{2} \cdot b \cdot h$.

☞ The measures of the angles of a triangle add up to $180°$.

☞ In any triangle, the sum of the measures of the angles is $180°$.

☞ If a rectangular solid (직육면체) has length l, width w, and height h, then the volume is given by $V = l \cdot w \cdot h$.

● circle (원)

☞ In any circle, if d is the diameter (지름) and r is the radius (반지름), then
$$d = 2 \cdot r.$$

☞ If C is the circumference (원주) of a circle and r is the radius, then
$$C = \pi d, \text{ or since } d = 2r, \ C = 2\pi r.$$

☞ The area of a circle of radius r is given by $A = \pi r^2$.

● $\approx$ means "approximately equal to"

● The inverse of any number x is named $-x$ (the inverse of x).

● The set of integers is made up of the whole numbers and all of their additive inverse.

☞ We call the natural numbers positive.

원서읽기를 위한 수학용어사전

☞ The inverses of the positive integers are called negative.
☞ Zero is not considered either positive nor negative.
☞ All negative integers are less than zero.
☞ All positive integers are greater than zero.

사칙연산에서 앞 수와 뒤 수의 명칭

연산	앞의 수	뒤의 수	결과
+	summand, addend 피가수	addend 가수	sum 합
−	minuend [mɪnjuènd] 피감수	subtrahend [sʌ́btrəhènd] 감수	difference 차
×	multiplicand [mʌ̀ltəplikǽnd] 피승수	multiplier 승수	product 곱
÷	dividend [dívidènd] 피젯수	divisor 제수	quotient & remainder 몫 & 나머지

● 정수의 절댓값은 수직선에서 원점으로 부터의 거리로 생각하면 된다.

: The absolute value of an integer can be thought of as its distance from 0 on the number line.

● absolute value

☞ The absolute value of 0 is 0.

☞ To add two negative integers, we add their absolute values and change the sign.

☞ We call $-a$ the additive inverse of a because adding

Chapter 2 영어교과서에 나타나는 수학용어

any number to its additive inverse always gives 0.

☞ To multiply a positive and a negative integer (number), multiply their absolute values. The answer is negative.

☞ To multiply two negative integers (numbers), multiply their absolute values. The answer is positive.

● When we divide a positive integer by a negative, or a negative integer by a positive, the answer is negative.

● When we divide two negative integers, the answer is positive.

● $-a$ is the inverse of a.

주의 $-a$ should not be read "negative a".

● To multiply several numbers, we can multiply two at a time. We can do this because multiplication of rational numbers is also commutative and associative.

● commutative (교환 가능한)
: 연산 *에 대해 $a*b = b*a$ 가 성립하는
● associative (결합 가능한)
: 연산 *에 대해 $(a*b)*c = a*(b*c)$ 가 성립하는
● 곱과 역수

☞ The product of three negative numbers is negative.

☞ The product of four negative numbers is positive.

☞ Two numbers are reciprocals of each other if their product is 1.

☞ Any nonzero number a has a reciprocal $\dfrac{1}{a}$.

☞ The reciprocal of a negative number is negative.

☞ The reciprocal of a positive number is positive.

☞ The reciprocal of $\dfrac{a}{b}$ is $\dfrac{b}{a}$.

☞ The reciprocal of $-\left(\dfrac{b}{a}\right)$ is $-\left(\dfrac{a}{b}\right)$.

덧셈의 역 : additive inverse

곱셈의 역 : reciprocal = multiplicative inverse

● rational numbers (유리수)

☞ The set of rational numbers (유리수) consists of all numbers that can be named with fractional notation $\dfrac{a}{b}$, where a and b are integers and b is not 0.

☞ A negative number divided by a positive number is negative.

☞ A positive number divided by a negative number is negative.

☞ For any rational numbers a, b and c,
$a(b-c) = ab - ac$.

Chapter 2 영어교과서에 나타나는 수학용어

(The distributive law (분배법칙) of multiplication over subtraction.)

● When all the terms of an expression have a factor in common, we can "factor it out" using the distributive laws.

● Factor out the largest common factor.
☞ The process of collecting like terms is also based on the distributive laws.
☞ $-1 \cdot a = -a$
: Negative one times a is the additive inverse of a.

☞ (괄호 앞에 마이너스 부호가 붙어 있을 경우) 괄호를 풀면 괄호 안의 모든 항의 부호를 + 는 - 로, - 는 + 로 바꾼다.
: Change the sign of every term in the parentheses.
$$3x - (4x + 2) = 3x + [-(4x + 2)]$$
Be sure to change the signs of all terms in parentheses(괄호).

● 음수를 짝수 번 곱하면 양수가 되고, 홀수 번 곱하면 음수가 된다.
: The product of an even number of negative number is positive. The product of an odd number of negative number is negative.

● The replacements that make an equation true are called solutions.

● To solve an equation means to find all its solutions.

● An equation $a = b$ says that a and b stand for the same number.

● The addition principle : If an equation $a = b$ is true, then $a + c = b + c$ is true for any number c.

● The multiplication principle : If an equation $a = b$ is true, then $a \cdot c = b \cdot c$ is true for any number c.

● 분사구문 형태의 수학 설명
☞ using a distributive law, ~ : 분배법칙을 사용하면 ~
☞ collecting like terms, ~ : 동류항을 모으면 ~
☞ adding 2, ~ : 2를 더하면 ~
☞ simplifying, ~ : 정리하며 ~
☞ multiplying by $\frac{1}{5}$, ~ : $\frac{1}{5}$을 곱하면 ~
☞ subtracting 5, ~ : 5를 빼면 ~
☞ diving by -720, ~ : -720 으로 나누면 ~
☞ adding exponents ~ : 지수를 더하면 ~
(예를들어 $3^3 \times 3^5$ 는 지수를 더하면 8이 되므로 3^8)

Chapter 2 영어교과서에 나타나는 수학용어

☞ adding an inverse ~ : 역수를 (양변에) 더하면 ~
☞ changing signs ~ : 부호를 바꾸면 ~
(양변에 −1을 곱하거나, 괄호 앞 음수 부호 때문에)

● cleared the fractions.
: 분수에 어떤 수(예를 들어 분모의 최소공배수)를 양변에 곱해서, 분모가 1이 되도록 하여, 정수가 되도록 함
● cleared the decimals.
: 소수에 어떤 수(예를 들어 10, 100, 1000 등)를 양변에 곱해서, 소수를 없애서, 정수가 되도록 함

● 방정식을 풀이하는 방법
$0.5x - 0.1x - 3 = 0.3x - 0.2x + 0.6$ 을 풀이하려면
: Solve for x, $0.5x - 0.1x - 3 = 0.3x - 0.2x + 0.6$.
① Multiply on both sides to clear of fractions or decimals.
: 양변에 10을 곱하여 분수나 소수를 없앤다.
$5x - x - 30 = 3x - 2x + 6$
② Collect like terms on each side, if necessary.
: 각 변에 동류항을 모은다.
$4x - 30 = x + 6$
③ Get all terms with variables on one side and all the other terms on the other side.
: 변수가 들어가는 항을 한 변으로 나머지를 다른 변으로 보낸다.
$4x - x = 6 + 30$

④ Collect like terms again, if necessary.
: 다시 동류항을 모은다.
   $3x = 36$
⑤ Divide to solve for the variable.
: 변수가 아닌 수로 나누어 변수값을 구한다.
   $x = 12$

● The Principle of Zero Product
: The product of two numbers is 0 if one of the numbers is 0.

● (서술형 문제를 수식화시키는 과정에서)
The first step in solving a problem is to translate it to mathematical language.

● 문제 풀이 단계 : Steps to use in solving a problem
Step 1. Translate to an equation.
Step 2. Solve the equation.
Step 3. Check the answer in the original problem.

● 초등수학에서 dimension 은 차원(次元)을 말하기 보다는 용적, 면적, 부피, 크기를 구하라는 것이다. 예를 들어
   Find the dimensions of the following solid figure.
   은 입체도형의 부피를 구하라이다.

Chapter 2 영어교과서에 나타나는 수학용어

● The measures of the angles of any triangle add up to $180°$.
: 삼각형의 내각의 합은 180도이다.

● 전압은 전류와 저항의 곱이다.
$$E = IR$$
(E : voltage, I : current, R : resistance)

● 속도는 거리를 시간으로 나눈 값이다.
$$\text{Speed} = \frac{\text{distance}}{\text{time}}.$$

● polynomial (다항식)
☞ evaluating polynomials : 다항식의 값을 구함
☞ Subtractions can be rewritten as additions.
 : 뺄셈은 (음수의) 덧셈으로 고쳐쓸 수 있다.

● 항과 인수의 차이
☞ terms (항) : things added
☞ factors (인수) : things multiplied

● Terms that have the same variable and the same exponent are called like terms or similar terms. (동류항)
☞ collecting like terms
☞ combining similar terms

● ascending order 와 descending order

☞ This polynomial is arranged in descending order. (내림차순)

$$8x^4 - 2x^3 + 5x^2 - x + 3$$

☞ The term with the largest exponent is first.

☞ The term with the next largest exponent is second, and so on.

☞ The opposite order is called ascending. (오름차순)

☞ The degree (차수) of a term is its exponent.

☞ The degree of a polynomial is its largest exponent.

주의 차(degree)와 계(order)

차(degree)는 다항식(polynomial)과 관련이 있으며 1차방정식, 2차방정식이라고 한다. 예를 들어 $x^2 - 3x + 7 = 0$ 을 2차방정식이라 한다. 이에 대하여 계(order)는 순서와 관련이 있으며 미분에서 1계도함수, 2계도함수라고 말하는데 이것은 미분을 한 번, 두 번 등과 같이 순서에 따라 미분한 횟수를 말한다.

● coefficients (계수)

☞ 계수가 0 이므로 나타나지 않는 항을 "missing term" 이라고 한다. 예를 들어 다항식 $x^3 - 2x + 3$ 에서 x^2 의 계수가 0 이므로 x^2 은 missing term (사라진 항, 보이지 않는 항)이다.

☞ If a coefficient is 0, we usually do not write the term.

☞ We say that we have a missing term. We leave spaces

Chapter 2 영어교과서에 나타나는 수학용어

for missing terms.

☞ Polynomials with just one term are called monomials.
: 항이 하나인 다항식을 단항식(monomial)이라고 한다.
☞ Polynomials with just two terms are called binomials.
: 항이 두 개인 다항식을 2항식(binomial)이라고 한다.
☞ Polynomials with just three terms are called trinomials.
: 항이 세 개인 다항식을 3항식(trinomial)이라고 한다.
☞ To add polynomials we can write a plus sign between them and collect like terms.
☞ If the sum of two polynomials is 0, they are additive inverses of each other.

● expand : (식이나 방정식을) 전개하다
☞ 예를 들어 6345를 expanded notation 으로 나타내면
$6 \cdot 10^3 + 3 \cdot 10^2 + 4 \cdot 10 + 5$

● FOIL 법칙 (앞밖안뒤 법칙)
두 다항식 $a+b$ 와 $c+d$ 의 곱은 한쪽의 항 각각에 대해 다른 쪽의 모든 항들과의 곱을 구한다. 즉
$(a+b)(c+d) = ac + ad + bc + bd$
이것을 일명 "FOIL 법칙"("앞밖안뒤 법칙")이라고 한다.
: To multiply two polynomials, multiply each term of one by every term of the other. Then add the results.

$$(a+b)(c+d) = ac + ad + bc + bd$$

① Multiply First terms (앞의 항끼리 곱한다)   : ac
② Multiply Outside terms (밖의 항끼리 곱한다) : ad
③ Multiply Inside terms (안의 항끼리 곱한다) : bc
④ Multiply Last terms (뒤의 항끼리 곱한다)   : bd

● Using $FOIL$, ~ : FOIL 법칙을 이용하면, ~

● 두 수식의 합과 차로 된 곱셈
: multiplying sums and differences of two expressions.
☞ The product of the sum and difference of two expressions is the square of the first expression minus the square of the second. 즉
$$(a+b)(a-b) = a^2 - b^2.$$

● 이항식의 제곱 : squaring binomials
☞ The square of a binomial is the square of the first expression, plus or minus twice the product of the two expressions, plus the square of the last. 즉
$$(a+b)^2 = a^2 + 2ab + b^2$$
$$(a-b)^2 = a^2 - 2ab + b^2$$

참고 (위와 같은 수식은) 기억하고 있지 않으면 나중에 (인수분해할 때) 어려움을 겪을 것이다.
  : You must memorize the rule. Otherwise, you

Chapter 2 영어교과서에 나타나는 수학용어

will have trouble later on.

● To factor (인수분해) an expression means to write it as a product. : 수식을 인수분해 한다는 것은 그것을 곱으로 나타내는 것을 의미한다.

자주 나타내는 표현은

☞ factoring out $4x^2$: $4x^2$ 으로 인수분해하면

☞ factoring $x^2 + 3x$

☞ factoring out the (a) common factor
: 공통인수로 인수분해하면

☞ separating into two binomials
: 두 개의 2항식으로 나누면

☞ factoring by grouping

☞ factoring out x again

● $x^2 + 5x + 6$ 을 인수분해 하려면 곱해서 6이 되고 합해서 5가 되는 두 수를 찾아야 한다.

: To factor $x^2 + 5x + 6$. We look for a pair of integers (or two numbers) whose product is 6 and whose sum is 5.

● 3항식을 인수분해하려면 시행착오(trial and error method) 방법을 이용해야 한다.

: When we factor trinomials we must use trial and error.

● (인수분해 하기 위해서는) 먼저 공통인수부터 찾아야 한다.
: First look for a common factor.

● A product is 0 iff (= if and only if) at least one of the factors is 0. (이것을 the principle of zero product 이라고 한다.)

● 답이 맞는가 검사한다.
: Check the answers in the problem.
● 문제를 식으로 나타낸다.
: Translate the problem to an equation.

● substitute A for B : B 대신 A 를 쓰다.
☞ Substituting 90 for N : N 대신 90 을 사용하다.
☞ squaring ~ : 제곱하면 ~
☞ Solving for y ~ : y 에 대해 풀면 ~

● Complete the square. : 완전제곱하다.
☞ 완전제곱(completing the square)은 이차항의 계수를 1이라 할 때, 일차항의 계수의 절반의 제곱을 더한 다음 다시 그 값을 빼준다. 예를 들어 $x^2 + 8x$ 일 때
taking half the x-coefficient : 4 (1차항 계수 8의 절반)
squaring : 16 (4 의 제곱)
이므로
$$x^2 + 8x = x^2 + 8x + 16 - 16 = (x+4)^2 - 16$$

Chapter 2 영어교과서에 나타나는 수학용어

● 좌표평면 (coordinate plane)

☞ x-axis

☞ y-axis

☞ ordered pairs : 순서쌍

☞ Plot the point $(-3, 4)$

: 좌표평면에 (-3, 4)의 위치를 기입하다.

☞ plot : 좌표로 위치를 결정하다.

☞ first coordinate : 첫 번째 좌표, x 좌표

☞ second coordinate : 두 번째 좌표, y 좌표

☞ the first quadrant : 제1사분면, 제1상한

☞ the second quadrant : 제2사분면, 제2상한

☞ In which quadrant is each point located?

☞ We substitute 3 for x and 7 for y.

: x 대신 3을, y 대신 7을 대입한다.

● 방정식의 그래프

☞ To graph an equation means to make a drawing of its solution.

☞ A graph of an equation is a picture of its solution set.

☞ The graph of $y = mx$ goes through the origin $(0, 0)$.

☞ The graph of any equation $y = mx + b$ is also a line. 여기서 $m =$ slope (기울기), $b = y$ −intercept (y −절편)

It is parallel to $y = mx$, but moved up or down.

☞ Any equation $y = mx + b$ has a graph that is a

straight line. It goes through the point $(0, b)$, the $y-$intercept, and has slope m.

☞ The x-intercept(절편) is in the form $(a, 0)$.
   To find a, let $y = 0$.
☞ The y-intercept is in the form $(0, b)$.
   To find b, let $x = 0$.
☞ The graph of $y = b$ is a horizontal line.
☞ The graph of $x = a$ is a vertical line.

● 연립방정식 (system of equations)
☞ A set of equations such as
$$\begin{cases} x + y = 8 \\ 2x - y = 1 \end{cases}$$
is called a system of equations. (연립방정식)
☞ If the lines are parallel, there is no solution.
☞ the substitution method : 대입법
☞ the addition method : 가감법
☞ Substituting 1 for x in the second equation.
  : 두 번째 방정식에 x 대신 1을 대입하면
☞ Multiply on both sides of the first equation by 5.
  : 첫 번째 방정식의 양변에 5를 곱한다.

● The degree (차수) of a term (항) is the sum of the exponents of the variables.

● The degree (차수) of a polynomial (다항식) is the

Chapter 2 영어교과서에 나타나는 수학용어

degree of the term of highest degree.
- Like terms (동류항, similar terms) have exactly the same variables with exactly the same exponents.
- The product of the sum and difference of two expressions is the square of the first expression minus the square of the second :
$$(a+b)(a-b) = a^2 - b^2$$
Factoring out the largest common factor.

● 분수식과 역

☞ Find the reciprocal of $\dfrac{2}{5}$.

☞ The reciprocal of $\dfrac{2}{5}$ is $\dfrac{5}{2}$.

☞ fractional expression : $\dfrac{3}{4}$, $\dfrac{5}{x+2}$, $\dfrac{x^2+3x-10}{7x^2-4}$

☞ To multiply two fractional expressions, multiply numerators and multiply denominators.

☞ When denominators are the same, we add the numerators and keep the denominator.

☞ When denominators are the same, we subtract the numerators and keep the denominator.

☞ To add when denominators are different we first find a common denominator.

☞ The numerator and denominator have no common

factor, so we cannot simplify.

☞ Always simplify at the end if possible.

☞ Don't forget parentheses.

● 다항식을 단항식으로 나누려면, 다항식의 각항을 단항식으로 나눈다.

: To divide by a monomial, we can divide each term by the monomial.

● 계수가 0인 항(missing term)은 빈자리로 남겨두어야 한다.(조립제법을 할 경우 특히 주의!)

: Leave space for missing terms.

● 최대공약수 : GCD (Greatest Common Divisor)
 최소공배수 : LCM (Least Common Multiple)

☞ A fractional equation (분수 방정식) is an equation containing one or more fractional expressions. To solve a fractional equation, multiply on both sides by the LCM (least common multiple : 최소공배수) of all the denominators (분모). This is called clearing of fractions. (분수를 없앤다고 한다. 분수를 제거한다고 한다.)

☞ Multiplying on both sides by the LCM, ~

☞ Multiplying on both sides by x, ~

☞ When clearing an equation of fractions, be sure to

Chapter 2 영어교과서에 나타나는 수학용어

multiply all terms in the equation by the LCM.

● It is important always to check when solving fractional equations. : (다른 문제의 경우와 같이) 분수 방정식을 풀고 난 후에 항상 검산하는 것이 중요하다.

● Removing parentheses ~ : 괄호를 없애면 ~

● If a job can be done in t hours, then $\frac{1}{t}$ of it can be done in 1 hour. : 어떤 일을 하는 데 t 시간이 걸린다면 1 시간 동안에는 $\frac{1}{t}$ 의 일을 하는 것이다.

ratio, rate, proportion 의 차이

☞ The ratio (비) of two quantities is their quotient.

　　2 : 3 → " 2 to 3 "

　　37% 는 the ratio of 37 to 100

☞ The ratio of two different kinds of measure is called a rate (비율).

☞ An equality of ratios, $\frac{A}{B} = \frac{C}{D}$, is called a proportion (비례).

● 지수와 근호

☞ What is the meaning of 5^2 ?

☞ The number c is a square root of a if $c^2 = a$.

☞ $\sqrt{a}$ ⋯ principal square root of a

☞ The symbol $\sqrt{}$ is called a radical symbol.

☞ radical expression 의 예

$$\sqrt{14}\ ,\ \sqrt{x}\ ,\ \sqrt{x^2+4}\ ,\ \sqrt{\frac{x^2-5}{2}}$$

☞ The expression written under the radical is called the radicand.

☞ $\sqrt{y^2-5}$: The radicand is y^2-5.

☞ Negative numbers do not have square roots.

☞ This is because the square of any negative number is positive. Thus the following expressions are meaningless : $\sqrt{-100}\ ,\ \sqrt{-49}\ ,\ -\sqrt{-3}$.

☞ Any radical expression $\sqrt{A^2}$ can be simplified to $|A|$.

● 무리수와 실수

☞ For any numbers a and b, $|a \cdot b| = |a| \cdot |b|$.

☞ An irrational number (무리수) is a number that cannot be named by fractional notation $\frac{a}{b}$, where a and b are integers and $b \neq 0$.

☞ Unless a whole number is a perfect square, its square root is irrational.

☞ π is irrational, and it is not a square root of any

Chapter 2 영어교과서에 나타나는 수학용어

rational number.

☞ The real numbers (실수) consist of the rational numbers and the irrational numbers. There is a real number for each point on a number line.

☞ For any nonnegative radicands A and B,
$\sqrt{A} \cdot \sqrt{B} = \sqrt{A \cdot B}$.

● 근사계산

☞ Rounded to three decimal places, $\sqrt{10} \approx 3.162$

$\sqrt{10}$ is approximately equal to 3.162.

: 소수점 아래 3자리까지 근사시키면

$\sqrt{10}$ 은 근사적으로 3.162와 같다.

☞ Rounded to 3 decimal places.
☞ Rounded to the nearest tenth.
☞ Rounded to three decimal places.

● 수식이 음수가 아니면 절댓값 기호가 필요하지 않다.

: Absolute value notation is not necessary because expressions are not negative.

● 분모를 유리화 한다.

: Rationalize the denominator

● 같은 근호란 근호 속의 숫자가 같다.
  (예를 들어 $\sqrt{3}$ 과 $2\sqrt{3}$ 은 like radical 이다.)
: Like radicals have the same radicands.

● For any nonnegative radicands A and B,
$$\frac{\sqrt{A}}{\sqrt{B}} = \sqrt{\frac{A}{B}}$$

● 직각삼각형은 3개의 변(side)을 가지고 있으며, 가장 긴 변을 빗변(hypotenuse)이라고 하고, 직각을 낀 두 변을 leg 라고 한다.
: In a right triangle, the longest side is called the hypotenuse ([haipɔ́tənjùːs] 빗변). The other two sides are called legs.

● The Pythagorean Property of Right Triangles :
In any right triangle if a and b are the lengths of the legs and c is the length of the hypotenuse, then
$$a^2 + b^2 = c^2.$$

● 직각삼각형에서 두변의 길이를 알면 나머지의 길이도 알 수 있다.
: In any right triangle, if we know the lengths of any two sides, we can find the length of the third side.

Chapter 2 영어교과서에 나타나는 수학용어

● The principle of squaring

☞ If an equation $a = b$ is true, then the equation $a^2 = b^2$ is true.

☞ 위의 역은 사실이 아니다.

: The converse (역) of the above is not true.

● 명제 : a proposition, a statement

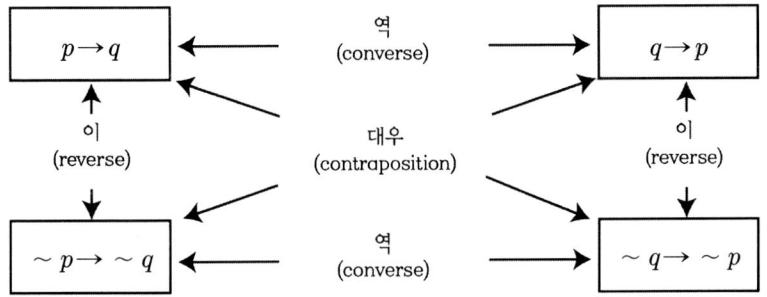

☞ 역 : converse

☞ 이 : reverse

☞ 대우 : contraposition

● It is important to check!

: 풀이 후 검산 과정이 필요!

● An equation of the type $ax^2 + bx + c = 0$, where a, b, and c are real number constants and $a > 0$, is called the standard form (표준형) of a quadratic equation (2차 방정식).

- $3x^2 + 2x - 4 = 0$. The equation is already in standard form.

- The only number with absolute value 0 is 0.
- Rationalizing the denominators ~ : 분모를 유리화하면 ~
- The length d of a diagonal of a rectangular solid (직육면체) is given by
$$d = \sqrt{a^2 + b^2 + c^2}$$
where a, b, and c are the lengths of the sides.

- $ax^2 + bx = 0$을 인수분해하여라.
: Factor $ax^2 + bx = 0$.
☞ A quadratic equation of this type will always have 0 as one solution and a nonzero number as the other solution.

- The number of diagonals (대각선), d, of a polygon (다각형) of n sides is given by
$$d = \frac{n^2 - 3n}{2}.$$
☞ An octagon has 8 sides. How many diagonals does it have?
: 팔각형은 8개의 변을 가졌는데, 대각선은 몇 개인가?

- Taking principal square root
$-2 \pm \sqrt{7}$ read [발음 red] "-2 plus or minus $\sqrt{7}$"

Chapter 2 영어교과서에 나타나는 수학용어

● We can make $x^2 + 10x$ the square of a binomial by adding the proper number to it. This is called completing the square (완전제곱).

● 완전제곱 관련

☞ Taking half the x -coefficient ~
 : 일차항계수(x 계수)의 절반을 취하면 ~

☞ Squaring ~ : 제곱을 하면 ~

☞ Adding ~ : 더하면 ~

☞ Compounding interest annually ~ : 년복리로 계산하면 ~

☞ Substituting into the formula ~ : 공식에 대입하면 ~

☞ Computing ~ : 계산하면 ~

☞ Taking the square root ~ : 제곱근을 취하면 ~

☞ Using table 1 ~ : 표 1을 이용하면 ~

☞ Completing the square ~ : 완전제곱하면 ~

☞ Multiplying by $\frac{1}{2}$ to make the $x^2 -$ coefficient 1 .

● 이차방정식의 근의 공식 (The Quadratic Formula) :

$$x = \frac{-b \pm \sqrt{b^2 - 4ac}}{2a}$$

☞ When $b^2 - 4ac \geq 0$, the equation has solutions.

☞ When $b^2 - 4ac < 0$, the equation has no real-number solutions.

☞ The expression $b^2 - 4ac$ is called the discriminant. (판별식)

☞ When using the quadratic formula, it is wise to compute the discriminant first. If it is negative, there are no solutions.

● 방정식 풀이
☞ Fractional Equations : 분수 방정식, 유리 방정식
☞ Radical Equations : 무리 방정식
☞ Adding -3 to get the radical alone on one side : 예를 들어 $\sqrt{x} + 3 = 7$ 일 때, 양변에 -3을 더한다.
☞ squaring both sides ~ : 양변을 제곱하면 ~
☞ Multiplying by a, to get h alone.
☞ Multiplying by g, to clear of fractions.
☞ Taking the square root ~
☞ Adding -1, to get y alone.
☞ Finding standard form ~
☞ Round to the nearest hundredth.
☞ Finding a common denominator ~

● 포물선 (parabola)
☞ line of symmetry parabola
☞ Some parabolas are thin and others are flat, but they all have the same general shape.

Chapter 2 영어교과서에 나타나는 수학용어

☞ Graphs of quadratic equations $y = ax^2 + bx + c$ are all parabolas. They are smooth cup-shaped symmetric curves, with no sharp points or kinks in them.

☞ The graph of $y = ax^2 + bx + c$ opens upward if $a > 0$. It opens downward if $a < 0$.

● 부등식 (inequality)

☞ An inequality is a number sentence with $>$, $<$, $\geq$, or $\leq$ for its verb.

☞ The letter x in $x < 2$ is called a variable.

☞ The replacement that make an inequality true are called its solutions. To solve an inequality means to find all of its solutions.

☞ If any number is added on both sides of a true inequality, we get another true inequality.

: 참인 부등식의 양변에 어떤 수를 더하면 또 다른 참인 부등식을 얻을 수 있다. 예를 들어 $5 > 2$ 가 참이므로 양변에 3을 더한 $5+3 > 2+3$ 역시 참이다.

● 부등식의 곱셈 법칙

: The Multiplication Principle for Inequalities

부등식의 양변에 양수를 곱하면 부등호가 변하지 않지만(The symbol stays the same.) 음수를 곱하면 $\leq$ 은 $\geq$ 로, $\geq$ 은 $\leq$ 로, $<$ 는 $>$ 로, $>$ 는 $<$ 로 바뀐다.(The symbol has to reversed.)

: If we multiply on both sides of a true inequality by a positive number, we get another true inequality. If we multiply by a negative number and the inequality symbol is reversed, we get another true inequality.

● 집합에서 조건제시법으로 나타난 수식 읽기

예를들어 $\left\{x \mid x \geq \dfrac{11}{12}\right\}$ 는

The set of all x such that x is greater than or equal to $\dfrac{11}{12}$.

로 읽히며, 각 부분을 다음과 같이 읽는다.

① { }   ⋯   The set of
② x ⋯   all x
③ | ⋯   such that
④ $x \geq \dfrac{11}{12}$ ⋯ x is greater than or equal to $\dfrac{11}{12}$

● $x \in A$ reads "x is a member of A." or "x belongs to A."

● The intersection (교집합) of two sets A and B is the set of members common to both sets and is indicated by the symbol $A \cap B$.

● Two sets A and B may be combined to form a new set. It contains the members of A as well as those of B.

Chapter 2 영어교과서에 나타나는 수학용어

The new set is called the union (합집합) of A and B, and is represented by the symbol $A \cup B$.

- 보각과 여각
  ☞ Supplementary angles(보각) are angles whose sum is $180°$.
  ☞ Complementary angles(여각) are angles whose sum is $90°$.

2.3 고급수학

- replace A by B : A를 B로 바꾸다
☞ Replacing % by $\times 0.01$, ~
: 퍼센트(%)를 $\dfrac{1}{100}$ ($\times 0.01$)로 바꾸면 ~

☞ repeating decimal : 순환소수
☞ repeating part : 순환소수에서의 순환부
☞ The sum of two negative numbers is negative. Add their absolute values and then make the answer negative.
☞ The reciprocal of a negative number is negative.

주의 $-(-2)$를 영어로 읽으면

① the inverse of negative 2. (맞음)
② the inverse of the inverse of 2. (맞음)
③ negative negative 2. (틀리게 읽음)
④ minus minus 2. (틀리게 읽음)

※ minus 는 '빼기'이고 negative 는 '음수'를 말한다.

결합법칙 (Associative law) : grouping has been changed.
교환법칙 (Commutative law) : order has been changed.

☞ Substituting ~ : 치환을 하면 ~
(cf) substitute A for B : B대신 A를 쓰다.
☞ Multiplying (on both sides by $\frac{1}{10}$) ~
: (양변에 $\frac{1}{10}$)을 곱하면 ~
☞ Adding ($4x$ on both sides) ~ : (양변에 $4x$ 를) 더하면 ~
☞ Writing p as a product of p and 1 . ($p = p \cdot 1$ 로 표시)
☞ Collecting like terms (on the left)
: (좌변에) 동류항을 모으면
☞ Combining like terms (on the left) ~
: (좌변에) 동류항을 결합하면 ~
☞ Using associativity ~ : 결합법칙을 이용하면 ~
☞ Changing the sign of every term
: 각 항의 부호를 변경하면
☞ Removing parentheses ~ : 괄호를 없애면
☞ Doing the calculations in the innermost parentheses ~
: 제일 안쪽에 있는 괄호안을 계산하면 ~
☞ We agree that a^0 means 1 . $(a \neq 0)$
: $a \neq 0$ 일 때 a^0 을 1 이라고 인정한다.

Chapter 2 영어교과서에 나타나는 수학용어

☞ By the definition of exponents ~
: 지수의 정의에 의해서 ~

☞ Simplifying ~ : 간단히 하면 ~

☞ Subtracting exponents : 지수끼리 빼면

☞ raise 3 to 5th power : 3을 5제곱하다

☞ To raise a power to a power we can multiply exponents.

$$(a^m)^n = a^{mn}$$

☞ Clear of fractions
: 분모의 최소공배수를 곱하여 분수를 없앤다

☞ Clear of decimals (if that is needed) 또는 (if necessary)
: (필요하다면) 양변에 10^n 을 곱하여 소수를 없앤다

☞ Multiplying to remove parentheses ~
: 괄호를 없애기 위해 곱해주면 ~

☞ We get a false equation. : 틀린 방정식을 얻는다.

☞ We get a true equation. : 옳은 방정식을 얻는다.

☞ Drawing a picture ~ : 그림을 그리면 ~

☞ translation : 해석, 평행이동

☞ Translate to an equation. : 방정식을 해석하다.

☞ Check the answer in the original problem.

● The measures of the angles of any triangle add up to 180°. : 삼각형의 세 내각의 합은 180도이다.

● $ax - b = cx$ 에서 Solve for x. (x 에 대해 풀어라.)
이면 먼저 x 가 들어가는 항을 좌변 (또는 우변)으로 모은다.
Collecting like terms, we have
$$(a-c)x = b$$
Dividing both sides by $(a-c)$
풀려는 변수를 남기고 양변 $a-c$ 로 나눈다.
We want this letter (x) alone.
$$x = \frac{b}{a-c}$$

● Factoring : 인수분해

직선방정식

(equation of straight lines)

☞ slope $= \dfrac{\text{rise}}{\text{run}}$: 기울기$= \dfrac{\text{수직으로의 변화}}{\text{수평으로의 변화}}$

☞ Think of 2 as $\dfrac{2}{1}$. : 2를 $\dfrac{2}{1}$ 으로 생각한다.

☞ Every horizontal line has a slope of 0.

☞ Vertical lines do not have a slope.

☞ $\approx$: is approximately equal to.

☞ m slope, (x_1, y_1) a point

Chapter 2 영어교과서에 나타나는 수학용어

The point-slope equation of a line is
$$y - y_1 = m(x - x_1).$$

☞ (x_1, y_1), (x_2, y_2) two points
The two-point equation of a nonvertical line is
$$y - y_1 = \frac{y_2 - y_1}{x_2 - x_1}(x - x_1).$$

☞ m slope, b y-intercept
The slope-intercept equation of a line is $y = mx + b$.

☞ a x-intercept, b y-intercept
The two-intercept equation of a line is $\dfrac{x}{a} + \dfrac{y}{b} = 1$.

☞ Two nonvertical lines are parallel if they have the same slope.

☞ Two lines are perpendicular if the product of their slopes is -1.

☞ Solving for y

☞ Fitting equations to data.

☞ Graph using intercepts.

☞ Translate to equations and solve.

☞ Equations have the same graph.

● consistent : 모순이 없는, 해를 (적어도 하나) 가지는

☞ If a system of equations has a solution, we say that it is consistent. 즉 They are consistent (모순없는, 해를 가지는) because there is at least one solution.

● inconsistent : 문제가 성립하지 않는, 해를 가지지 않는

☞ If a system does not have a solution, we way that it is inconsistent.

☞ Annual Percentage Rate (APR)
: 매년 지불 할 총 대출 금액에 대한이자 금액

$$APR = \frac{Total\ Interest}{Average\ Principal}$$

☞ It is illegal in the United States for a lender not to inform you of the APR.

☞ Substituting $y + 1$ for x,

☞ Adding -1 on both sides,

☞ Multiplying on both sides by $\frac{1}{2}$,

☞ Multiplying to remove parentheses.

● 연립방정식을 풀이하는 방법으로 3가지 방법을 이용하지만, 미국 교과서에서는 등치법을 별도의 방법으로 구분하여 설명하고 있지는 않다.

① the addition method. (가감법)

Chapter 2 영어교과서에 나타나는 수학용어

② the substitution method. (대입법)
③ the equivalence method. (등치법)

● Multiplying (= When we multiply) by -2 on both sides, ~ : 양변에 –2를 곱하면 ~
● Substituting 2 for x in second equation,
: 두 번째 식에다 x 대신 2를 대입하면
● Multiplying on both sides by 10,
: 양변(좌변과 우변)에 10을 곱하면
● Multiplying in the first equation by -2,
: 첫 번째 식에다 –2를 곱하면
● Clear of decimals.
: (양변 또는 분자분모에 적당한 상수배를 하여) 소수를 없앤다.
● Clear of fractions.
: (분모의 최대공약수를 곱하여) 분수를 없앤다.
● Multiplying to remove parentheses on the left,
: 좌변에 괄호를 없애기 위해 (적당한 수를) 곱하면

● 3원 3차 방정식 (three equations in three unknowns)
① We first use any two of the three equations to get an equation in two variables.
② We use a different pair of equations and eliminate the same variable that we did in (①).

● The sum of the measures of the angles is 180° in any triangle.
● The value of a determinant (행렬식) is a number.
: 행렬식의 값은 (벡터가 아닌) 수이다.

● break-even point : 손익분기점
● equilibrium point : 평형점
● Price in dollars. : 달러(로 나타낸) 가격

● The magic number in baseball (야구에서 매직넘버) can be found by using the polynomial $G - P - L + 1$. Here G is the number of games in the season, P is the number of games the leading team has played, and L is the number of games ahead in the loss column.

● Arrange $-2x + x^2 + 5x^7 + 6x^5$ in descending order.
　　　　　　　　　　　　　　　　　in ascending order.
　　　　　　　　　　　　　　　　　in descending powers of y.
　　　　　　　　　　　　　　　　　in ascending powers of y.
● 다항식 : Polynomials
☞ Polynomials with just one term are called monomials. (단항식)
☞ Polynomials with just two terms are called binomials. (이항식)
☞ Those with just three terms are called trinomials. (삼항식)

● 다항식끼리의 뺄셈의 경우 뒤 다항식의 부호가 변경된다.

Chapter 2 영어교과서에 나타나는 수학용어

: Don't forget to change the sign of every term.
: Changing the sign.

● To multiply two polynomials, multiply each term of one by every term of the other. Then add the results, combining like terms (동류항) if possible.

$$(A + B)(C + D) = AC + AD + BC + BD$$

$$F \quad\ \ O \quad\ \ I \quad\ \ L$$

first   outside   inside   last
terms   terms   terms   terms

● Combining like terms : 동류항을 결합하면

☞ $(A + B)^2 = A^2 + 2AB + B^2$
☞ $(A - B)^2 = A^2 - 2AB + B^2$

: The square of a binomial is the square of the first expression, plus or minus twice the product of the expressions, plus the square of the second expression.

☞ $(A + B)(A - B) = A^2 - B^2$

: The product of the sum and difference of two expressions is the square of the first expression minus the square of the second.

● Factors (인수) are what we can multiply together to get

an expression.

- factor out : 인수분해하다

☞ Factoring out an x, : x 로 인수분해하면

☞ Factoring out an x again, : 다시 x 로 인수분해하면

☞ Factoring out another x, ~

☞ Always factor out the largest factor common to the terms. : 항상 항 중에서 가장 큰 공통인수로 인수분해한다.

☞ Factoring the difference of squares,

☞ Removing the common factor,

☞ Always continue to factor as long as you can.

☞ That way you will be factoring completely.

- n 개의 변을 가지는 다각형의 대각선의 개수

: The number of diagonals of a polygon having n sides is given by the polynomial $\frac{1}{2}n^2 - \frac{3}{2}n$.

☞ Find the number of diagonals of a polygon with 5 sides. : 5개의 변을 가지는 다각형의 대각선의 수를 구하여라.

- 3세개의 항으로 되는 제곱식 : trinomial squares

$$A^2 \pm 2AB + B^2 = (A \pm B)^2$$

☞ Removing the common factor,

☞ Always remember to look first for a common factor.

Chapter 2 영어교과서에 나타나는 수학용어

- FOIL 법칙

$(a+b)(c+d) = ac + ad + bc + bd$

$\quad\quad\quad\quad\quad\quad F \quad O \quad I \quad L$

first outside inside last

- $A^2 + B^2$ cannot be factored in real numbers.
: 실수범위 안에서 인수분해되지 않는다.
☞ factor over real numbers : 실수 범위까지 인수분해하여라.
☞ factor over complex numbers

> [참고] 약수와 배수는 자연수, 정수와 다항식에서만 생각할 수 있다. 다항식의 계수(coefficient)의 집합을 어느 범위로 한정하느냐에 따라 인수분해의 결과가 달라질 수 있다. 계수를 정수 또는 유리수로 한정하는 경우를 공식으로 주로 사용한다.

- Difference of squares : $A^2 - B^2 = (A+B)(A-B)$
- Sum of cubes : $A^3 + B^3 = (A+B)(A^2 - AB + B^2)$
- Difference of cubes

$\quad\quad : A^3 - B^3 = (A-B)(A^2 + AB + B^2)$

[인수분해 할 때]

① Try trial and error (시행착오 방법).
② Factoring out the largest common factor

factoring two grouped binomials : $3(x+4) + ax(x+4)$.
Solve for $b^2 - 2b = 0$.
Factoring, we have $b(b-2) = 0$.
Using the principle of zero products, $b = 0$ or $b = 2$.

- $\dfrac{x}{a^2} = 5$ 일 때, x 에 대해서 풀려면

: $\dfrac{x}{a^2} = 5$. Solve for x.

양변에 a^2 을 곱해서 좌변에 x 만 남게한다.
: Multiplying by a^2 to get x alone.

- Simplify $\dfrac{a}{b} \cdot \dfrac{c}{d}$.

: Multiplying numerators and multiplying denominators, we have $\dfrac{a}{b} \cdot \dfrac{c}{d} = \dfrac{ac}{bd}$.

- Factoring numerator and denominator.

- A complex fractional expression (복소분수식이 아니라 번분수식(繁分數式)이라는 의미) is one that has a fractional expression in its numerator or its denominator, or both. These are complex fractional expressions :

Chapter 2 영어교과서에 나타나는 수학용어

$$\frac{x}{x-\dfrac{1}{3}},\ \frac{\dfrac{5}{x}}{\dfrac{x}{y}},\ \frac{\dfrac{1}{a}+\dfrac{1}{b}}{\dfrac{1}{a}-\dfrac{1}{b}}$$

● To divide a polynomial by a monomial we can divide each term by the monomial.

● 피젯수 ÷ 제수 = 몫 ⋯ 나머지
 : Dividend ÷ Divisor = Quotient ⋯ Remainder

$$\begin{array}{r} x\phantom{{}+5x+8} \\ x+3\ \overline{\smash{\big)}\ x^2+5x+8} \\ \underline{x^2+3x} \\ 2x+8 \end{array}$$

↑
다음 나눗셈을 수행하기 위해 8은 그대로 내려온다.
 : The 8 has been "brought down (내려온다)".

● long division (장제법)과 단제법 (short division)
☞ 보통 우리가 나눗셈을 할 때 사용하는 방법이 long division 이고, 이것의 간략한 형태가 short division 이다. short division 은 주로 제수(divisor)가 한자리 수일 때 암산과 병행하여 사용한다.
 : The abbreviated form of long division is called short division, which is almost always used instead of long division

when the divisor has only one digit.

● synthetic division (조립제법)
☞ 계수만 적어서 하는 나눗셈(조립제법, 조립나눗셈)을 말하며 다항식의 근을 구할 때 주로 사용한다.
: Synthetic division is a shorthand, or shortcut, method of polynomial division in the special case of dividing by a linear factor -- and it only works in this case. Synthetic division is generally used, however, not for dividing out factors but for finding zeroes (or roots) of polynomials.

☞ $(4x^3 - 3x^2 + x + 7) \div (x - 2)$ 을 조립제법으로 구하려면

```
         4   -3    1    7
    2  |      8   10   22
       ─────────────────────
  →      4    5   11   29
```

계수만 적어서 계산을 하며 더해서 내려오고, 곱해서 올라가는 계산을 반복한다.

☞ missing term : 계수가 0 인 항. 예를 들어 $x^3 + 2x - 3$ 에서 x^2 은 missing term 이다.

● 분수방정식의 양변에 분모의 최소공배수를 곱해서 분모를 없애주는 것을 cleared the fraction 이라고 한다.
: To solve a fractional equation we multiply on both sides by the LCM of all the denominators. This is called clearing

Chapter 2 영어교과서에 나타나는 수학용어

of fractions.

☞ Multiplying on both sides by the LCM, (we have) ~
☞ Multiplying to remove parentheses, ~
☞ Multiplying and collecting like terms, ~
☞ Using the principle of zero products, ~
☞ Getting 0 on one side, ~
: 한 변(좌변 또는 우변) 0 이 되도록 하려면 ~
☞ Factoring, ~
☞ Simplifying, ~

- If a job can be done in t hours, then $\dfrac{1}{t}$ of it can be done in 1 hour.
- 속도 $= \dfrac{거리}{시간}$: speed $= \dfrac{\text{distance}}{\text{time}}$ $(r = \dfrac{d}{t})$
- Factoring out r, ~
- Multiply if necessary.
- Factoring out the unknown, ~ : 미지수로 인수분해 하면
- $y = kx$ 는 "y varies directly as x." 이다. 이때 k is called the variation constant.
- 지수법칙

☞ $a^m a^n = a^{m+n}$: Add exponents when multiplying
☞ $\dfrac{a^m}{a^n} = a^{m-n}$: Subtract exponents when dividing

☞ $(a^m)^n = a^{mn}$: Multiply exponents when raising a power to a power.

☞ a^n and a^{-n} are reciprocals.

☞ $a^1 = a$

☞ $a^0 = 1$ if $a \neq 0$

● scientific notation (과학적 표기법)

: $a \times 10^b$ or 10^b, $1 < a < 10$

☞ Converting 0.8 to scientific notation, ~

: 0.8을 과학적 표기법으로 바꾸면 ~

● 소수점 이동 - (적당한 10의 지수를 곱하면)

☞ Moving decimal point 5 places to the right,

☞ Moving decimal point 8 places to the left,

● compound interest (복리 이자)

$$A = p\left(1 + \frac{r}{n}\right)^{nt}$$

여기서

A : amount of money (원리합계)

p : principal (원금)

r : annual interest rate (연이율)

n : number of times per year interest is compounded

t : the time in years.

Chapter 2 영어교과서에 나타나는 수학용어

참고 은행의 예금 상품은 연이율로 제시된다. 1년에 이자 계산을 n 번하는 복리 예금의 경우 매번 $\dfrac{\text{연이율}}{n}$ 의 이율로 이자를 계산한다.

이때, $\dfrac{(1\text{년 후의 이자총액})}{(\text{원금})} \times 100\,(\%)$ 을 실효수익율(effective interest rate)이라고 한다.

● A square root (제곱근) of a number a is a number c whose second power is a, that is, $c^2 = a$.

☞　5 is a square root of 25.
☞　-5 is a square root of 25.

● 제곱근 (square root)
☞ Every positive real number has two real-number square roots.
☞ The number 0 has just one square root, 0 itself.
☞ Negative numbers do not have real number square roots.
☞ The symbol $\sqrt{}$ is called a radical (근).
☞ $\sqrt{5}$: 5 is called a radicand (근호 속의 숫자).
☞ The principal (nonnegative) square root of a^2 is the absolute value of a.
☞ $\sqrt{a^2} = |a|$, for all real a.

● 세제곱근 (cube root)　주의　cubic root 가 아니다.

☞ Every real number has exactly one cube root in the system of real numbers.

☞ The cube root of a positive number is positive.

☞ The cube root of a negative number is negative.

☞ The cube of 4 is 64. 즉 $4^3 = 64$.

☞ The cube root of 64 is 4. 즉 $\sqrt[3]{64} = 4$.

● n 제곱근 (the n-th root)

☞ $\sqrt[k]{}$ 에서 k를 index 라고 한다.

☞ 제곱근일 경우는 $\sqrt[2]{}$ 로 쓰지 않고 $\sqrt{}$ 로 적는다.

: When the index is 2 we do not write it.

☞ for all real (실수) a

(1) $\sqrt[k]{a^k} = |a|$ when k is even.

(2) $\sqrt[k]{a^k} = a$ when k is odd.

● Factoring into two radicals,

☞ Factoring the radicand, $\sqrt{5x^2} = \sqrt{x^2 \cdot 5}$.

☞ Taking the square root of x^2, ~

☞ $\dfrac{5}{16} = 0.3125$ → ending (terminating) decimal.

☞ $\dfrac{3}{11} = 0.2727\cdots = 0.\overline{27}$ → repeating decimal.

Chapter 2 영어교과서에 나타나는 수학용어

☞ $\pi = 3.14159265 \cdots$ 무리수 (Numeral does not repeat.)

- Collecting like radical terms, ~

- Squaring a binomial ~
$$(a + b)^2$$
- Using FOIL, ~
- Renaming 4 as 2^2, ~
- $\sqrt{x} - 3 = 4 \Rightarrow \sqrt{x} = 7$

: Adding 3 to isolate the radical, ~

- $\sqrt{x-3} + \sqrt{x+5} = 4 \Rightarrow \sqrt{x-3} = 4 - \sqrt{x+5}$

: Adding $-\sqrt{x+5}$; this isolates one of the radical terms.

- Isolating the remaining radical term, ~
- We square both sides, ~
- Squaring both sides, ~
- Factoring, ~
- Using the principle of zero products, ~
- Using a distributive law, ~
- The conjugate (켤레, 공액) of $a - bi$ is $a + bi$.
- Multiplying by 1, ~
- Equations of the second degree are called quadratic (2차의).

- 2차방정식의 근의 공식 : the quadratic formula

☞ 2차라는 말이 있으면 특별히 $a \neq 0$ 이라고 따로 적지 않는다.

☞ The solutions of a quadratic equation (이차방정식) $ax^2 + bx + c = 0$ are given by

$$x = \frac{-b \pm \sqrt{b^2 - 4ac}}{2a}.$$

☞ 이차방정식의 판별식 : Discriminant of a Quadratic

판별식 $D = b^2 - 4ac$ 은 이차방정식(quadratic equation)의 근이 어떤 형태인지를 알려준다.

: The number $D = b^2 - 4ac$ determined from the coefficients of the equation $ax^2 + bx + c = 0$. The discriminant reveals what type of roots the equation has.

- Taking square roots, ~
- Finding standard form, ~
- Rearranging, ~
- Rounding to the nearest tenth, ~
- Using a distributive law, ~
- Writing standard form, ~
- Using the principle of zero products, ~
- Multiplying by the LCM, ~

Chapter 2 영어교과서에 나타나는 수학용어

● 왕복여행에서 trip out 과 return trip

    Trip out   Return trip

    $A \to B$ $A \leftarrow B$

● Getting 0's on one side, : 좌변 또는 우변을 0을 만들면, 예를 들어 $x = \sqrt{3} \ \to \ x - \sqrt{3} = 0$.

● Clearing of fractions,
● Adding $-b^2$ to get a^2 alone, ~
● Writing the equation in standard form, ~
● Using the quadratic formula, ~
● Getting all a^2 terms together, ~
● Factoring out a^2, ~
● Taking the square root, ~
● Thinking of x^4 as $(x^2)^2$, ~

● 정비례, 반비례, 연합하여 비례

☞ y varies directly as the square of x if there is some positive constant k such that $y = k x^2$.

  : vary directly (정비례하여 변화하다)

☞ y varies inversely as the square of x if there is some positive constant k such that $y = \dfrac{k}{x^2}$.

  : vary inversely (반비례하여 변화하다)

☞ y varies jointly as x and z if there is some positive constant k such that $y = kxz$.

: varies jointly (연합하여 변화하다, 함께 변화하다)

● 포물선을 그릴 때 주의할 점

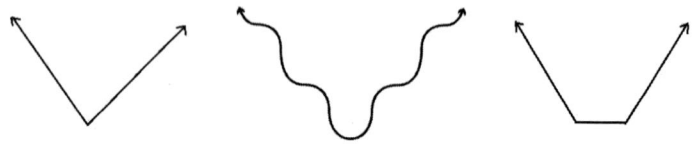

(1) Sharp point is wrong.   (2) Kinks are wrong.   (3) Flat nose is wrong.

(1) 꼭짓점을 뾰족하게 하면 안 되고,
(2) 꼽슬꼽슬하게 해서도 안 되며,
(3) 꼭짓점을 평편하게 해서도 안 된다.

☞ parabola : 포물선
☞ line of symmetry : 대칭축
☞ vertex : 꼭짓점

● 피타고라스 정리 : The Pythagorean Theorem

: In a right triangle, if c is the length of the hypotenuse (빗변) and a and b are the lengths of the other two sides (the legs), then $a^2 + b^2 = c^2$.

Chapter 2 영어교과서에 나타나는 수학용어

● 거리 공식 : The Distance Formula

: The distance between any two points (x_1, y_1) and (x_2, y_2) is given by

$$d = \sqrt{(x_1 - x_2)^2 + (y_1 - y_2)^2}\ .$$

● Equation of a circle (원의 방정식)

☞ A circle (원) with center (a, b) and radius r has an equation

$$(x - a)^2 + (y - b)^2 = r^2 \qquad \text{(Standard form)}$$

☞ A circle centered at the origin with radius r has an equation

$$x^2 + y^2 = r^2\ .$$

● The ellipse (타원) is defined to be the set of all points in a plane such that the sum of the distances from two fixed points F_1 and F_2 (called foci)(초점) is constant.

● Equation of an ellipse (타원의 방정식)

☞ An ellipse with its center at the origin has an equation

$$\frac{x^2}{a^2} + \frac{y^2}{b^2} = 1 \quad (a,\ b > 0) \quad \text{(standard form)}.$$

☞ Planets and comets have orbits around the sun that are ellipses. The sun is located at one focus.

● Equation of an hyperbola (쌍곡선의 방정식)

☞ Hyperbolas with their centers at the origin have equations as follows.

$$\frac{x^2}{a^2} - \frac{y^2}{b^2} = 1 \quad \text{(Axis horizontal)}$$

$$\frac{y^2}{b^2} - \frac{x^2}{a^2} = 1 \quad \text{(Axis vertical)}$$

● asymptotes of a hyperbola (쌍곡선의 접근선)

☞ asymptote 또는 asymptotic line : 점근선(漸近線)

☞ For hyperbolas with equations as given above, the asymptotes are the lines

$$y = \frac{b}{a}x \quad \text{and} \quad y = -\frac{b}{a}x.$$

☞ Hyperbolas having the $x-$ and $y-$ axis as asymptotes have equations as follows : $xy = c$, where c is a nonzero constant.

● Letting $u = x^2$, ~

● This curve does not touch or cross the $y-$ axis.

Chapter 2 영어교과서에 나타나는 수학용어

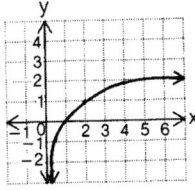

$$y = \log_2 x$$

- exponential function (지수 함수)
☞ Converting to an exponential equation, ~

- logarithmic function (로그 함수)
☞ $y = \log_a x$ is defined to mean $x = a^y$, where $x > 0$ and a is a positive constant different from 1.

☞ $\quad \log_a MN = \log_a M + \log_a N$.

: The logarithm of a product is the sum of the logarithms of the factors.

☞ $\quad \log_a M^p = p \cdot \log_a M$.

: The logarithm of a power of M is the exponent times the logarithm of M.

☞ $\quad \log_a \dfrac{M}{N} = \log_a M - \log_a N$.

: The logarithm of a quotient is the logarithm of the dividend minus the logarithm of the divisor.

☞ $\log_a a^k = k$ for any number k

☞ $\log 5240 \approx 0.7193 + 3$ (지표+가수)

가수 : mantissa (a number from 0 to 1)

지표 : characteristic (an integer)

- Solving ~
- Writing exponential notation, ~
- Rounding to four decimal places, ~

- Taking the antilog, ~ : 로그의 역(진수)을 취하면, ~
- Taking log on both sides, ~ : 양변에 로그를 취하면, ~
- Taking the antilog on both sides, ~
- Multiplying by (-1) and reversing, the inequality sign, $-x < -3 \rightarrow x > 3$.

☞ plane geometry (평면기하학)
☞ solid geometry (입체기하학)

- 집합의 표현과 읽기

$\{x \mid x \leq \sqrt{2}\}$는 "The set of all x such that x is less than or equal to $\sqrt{2}$."

이들 각각은

$\{\ \}$	$\cdots$	The set of
x	$\cdots$	all x
$\mid$	$\cdots$	such that
$x \leq \sqrt{2}$	$\cdots$	x is less than or equal to $\sqrt{2}$.

Chapter 2 영어교과서에 나타나는 수학용어

☞ decomposition into partial fractions

주의 비(ratio)와 비율(rate)의 차이

비(ratio)란 단위가 같은 경우 나눗셈을 할 때 나타나고, 비율(rate) 또는 변화율(rate of change)은 단위가 다른 경우 나눗셈에서 나타난다.

: A ratio is the quotient of two quantities which have the same unit. (Note that the units are not written.)

: A rate is the quotient of two quantities which have different units. (The units are written as part of the rate.)

☞ Please note the difference between ratio, simplest rate, and unit rate.

For the problem: "24 children in 10 families"

① ratio : $\dfrac{12}{5}$ (no units)

② simplest rate : $\dfrac{12\ children}{5\ families}$

  (units required; denominator not necessarily = 1)

③ unit rate : 2.5 children per family

  (units required; denominator = 1)

주의 영어에서 'between a and b'는 a와 b를 포함하는 것(inclusive sense)으로 이해하여야 한다. 예를 들어 'integers between 1 and 10'은 $1, 2, 3, \cdots, 10$을 의미한다. 이 경우 1과 10이 제외되어야 한다면 between a and b exclusive 로 표시한다.

Chapter 3 수의 명칭

Chapter 3 수의 명칭

이 장에서는 주로 수에 관련된 명칭에 대해서 언급하기로 한다. 철자에 특히 주의해야할 곳에는 ★표를 붙였다.

3.1 우리나라 수의 명칭

우리나라는 중국, 일본과 같은 단위명칭을 사용하는데, 큰 수인 경우에는 같지만 작은 수인 경우에는 우리나라와 일본은 같은 단위명칭이지만, 중국에서는 할(割)이 없이 하나씩 올라가면서 사용된다.

중국어 수사		일본어 수사	
百(백)	bǎi [빠이]	百(백)	yaku(ゃく)
千(천)	qiān [치앤]	千(천)	sen (せん)
万(만)	wn [완]	万(만)	man(まん)
亿(억)	y [이]	億(억)	oku(おく)
兆(조)	zho [자오]	兆(조)	tyou(ちょう)
京(경)	jīng [징]	京(경)	kei(けい)

3.1.1. 큰 수

일(一), 이(二), 삼(三), 사(四), 오(五), 육(六), 칠(七), 팔(八), 구(九), 십(十) 의 열 개의 기본에 더하여 각 자리를 열 배하면 일(一), 십(十), 백(百), 천(千), 만(萬)이 되고, 일(一)부터 계속 그것을 만(萬)배하면 다음이 얻어진다. (참조 : 인도 불교경정 화엄경)

단위명칭	크기	비 고
만(萬)	10^4	
억(億)	10^8	
조(兆)	10^{12}	
경(京)	10^{16}	
해(垓)	10^{20}	
자(秭)	10^{24}	"시(秭)"라고 적힌 문헌은 한자를 혼돈한 것으로 생각
양(穰)	10^{28}	
구(溝)	10^{32}	
간(澗)	10^{36}	
정(正)	10^{40}	
재(載)	10^{44}	천재(千載)는 10^{47} 이고 천재일우(千載一遇)는 확률 $\frac{1}{10^{47}}$ 을 말한다.
극(極)	10^{48}	('정도가 심한'이란 뜻으로 사용) 극빈자, 극존칭, 극초단파
항하사 (恒河沙)	10^{52}	또는 恒河砂 (인도 갠지스강의 모래라는 뜻)
아승기 (阿僧祇)	10^{56}	(asamkhya)
나유타 (那由他)	10^{60}	
불가사의 (不可思議)	10^{64}	(헤아릴 수 없을 정도로 많다)
무량대수 (無量大數)	10^{68}	(아미타불과 그 땅에 사는 백성들의 수명이 끝없음)

3.1.2. 작은 수

두산동아 '새국어사전'에는 중국식 기준으로 나타나내고 있으며, 우리나라와 일본에서는 아래 단위에 적힌 지수가 하나씩 감소한다. 예를 들어 모(毛)일 경우 중국에서 10^{-3}이 우리나라와 일본에서는 10^{-4}이 된다.

중국에서는 할(割)을 사용하지 않고 우리나라와 일본에서는 할(割)을 사용하기 때문에 아래 표를 이용할 경우 주의가 필요하다.

단위명칭	크기	비 고
할(割)		중국에서는 사용안 함 우리나라와 일본에서는 10^{-1}
푼(분(分))	중국에서 10^{-1}	우리나라와 일본에서는 10^{-2} - 이하 중국식기준으로 작성 -
리(厘 또는 釐)	푼의 1/10 , 10^{-2}	털
모(毛)	10^{-3}	실
사(絲)	10^{-4}	
홀(忽)	10^{-5}	작다　〈참고〉 홀연(忽然)
미(微)	10^{-6}	가늘다
섬(纖)	10^{-7}	

원서읽기를 위한 수학용어사전

(표 계속)

단위명칭	크기	비 고
사(沙 또는 砂)	10^{-8}	
진(塵)	10^{-9}	먼지
애(埃)	10^{-10}	티끌
묘(渺)	10^{-11}	아득함 〈참고〉 묘연(渺然)
막(漠)	10^{-12}	〈참고〉 막연(漠然)
모호(模糊)	10^{-13}	
준순(逡巡)	10^{-14}	
수유(須臾)	10^{-15}	잠시 동안
순식(瞬息)	10^{-16}	순식간의 준말
탄지(彈指)	10^{-17}	
찰나(刹那)	10^{-18}	눈 깜짝할 사이에는 3000 번의 찰나가 있다.
육덕(六德)	10^{-19}	
허공(虛空)	10^{-20}	텅 빈 공중
청정(淸淨)	10^{-21}	먼지가 전혀 없는 깨끗한 상태

Chapter 3 수의 명칭

3.2 영어 관련 수사

미국에서 쓰는 수의 단위는 4자리 이상의 숫자는 뒤에서 세 개씩 나타나도록 읽기 쉽도록 comma 를 찍는다.

영어로 큰 수 읽기

10^3	thousand	1천
10^6	million	백만
10^9	billion	십억
10^{12}	trillion	1조
10^{15}	quadrillion	천조
10^{18}	quintillion	백경
10^{21}	sexillion	십해
10^{24}	septillion	1자
10^{27}	octillion	천자
10^{30}	nonillion	백양
10^{33}	decillion	십구
10^{36}	undecillion	1간
10^{39}	duodecillion	천간
10^{42}	tredecillion	백정
10^{45}	quattuordecillion	십재
10^{48}	quindecillion	1극

원서읽기를 위한 수학용어사전

10^{51}	sexdecillion	천극
10^{54}	septdecillion	백항아사
10^{57}	octodecillion	십아승기
10^{60}	novemdecillion	1나유타
10^{63}	vigintillion	천나유타
10^{66}	unvigintillion	백불가사의
10^{69}	duovigintillion	십무량대수
10^{72}	trevigintillion	
10^{75}	quattuorvigintillion	
10^{78}	quinvigintillion	
10^{81}	sexvigintillion	
10^{84}	septvigintillion	
10^{87}	octovigintillion	
10^{90}	novemvigintillion	
10^{93}	trigintillion	
10^{96}	untrigintillion	
10^{99}	duotrigintillion	
10^{100}	google, 구골(googol)	원래 구골(googol)이었으나 오타 혹은 오류로 google 로 됨
10^{google}	googleplex	
	zillion(미국)	터무니없이 큰 수 (million과 billion을 모방)
	stil·lion [stíljən]	(미·속어) 엄청나게 큰 수

Chapter 3 수의 명칭

> **참고** 사실상 현실생활에서는 구골보다 큰 수는 없다. 지구상에 있는 전체 모래의 개수도 10^{30} 을 넘지 않는다. 은하에 있는 별의 수도, 지구상에 사는 인간의 모든 머리카락의 수의 합도 10^{20} 을 넘지 않는다.

3.3 특수한 수사

1/2	half
1	single
2	couple
6	a half dozen, half a dozen
12	dozen
18	a dozen and a half
20	score
2배	double
3배	triple, treble
4배	quadruple
1번	once
2번	twice
3번	three times, thrice
10년	decade
100년	millennium
10억년	aeon (무한히 긴 시대를 의미)

3.4 주의할 영어의 기수와 서수

	기수	서수	비고
1	one	first★	
2	two	second★	
3	three	third★	
4	four	fourth	
5	five	fifth★	5와 12 처럼 ve 로 끝나면 f 로 바꾸고 th를 붙인다.
8	eight	eighth★	t를 한번만 적는다.
9	nine	ninth★	e를 생략한다.
10	ten	tenth	
12	twelve	twelfth★ 또는 dozenth★	
20	twenty	twentieth★	
21	twenty-one	twenty first★	
30	thirty	thirtieth★	
100	one hundred	(one) hundredth	
1000	one thousand	(one) thousandth	
n	n	n -th	
$n+1$	$n+1$	$n+1$ st	$n+2$ nd, $n+3$ rd, $n+4$ th

참고 20, 30, 40, 50, ⋯ , 90 은 맨 끝의 y 를 i 로 바꾼 다음 e 를 붙이고 나서 th 를 붙인다.

Chapter 3 수의 명칭

3.5 그리스어 라틴어 관련 수사

배수와 약수의 접두사

그리스어			라틴어			켈트어		
접두어	인자	기호	접두어	인자	기호	접두어	인자	기호
엑사 (exa-)	$10^{18}, 2^{60}$	E	데시 (deci-)	10^{-1}	d	펨토 (femto)	10^{-15}	f
페타 (peta-)	$10^{15}, 2^{50}$	P	센티 (centi-)	10^{-2}	c	아토 (atto)	10^{-18}	a
테라 (tera-)	$10^{12}, 2^{40}$	T	밀리 (milli-)	10^{-3}	m			
기가 (giga-)	$10^{9}, 2^{30}$	G	마이크로 (micro-)	10^{-6}	μ	더 큰 수와 더 작은 수		
메가 (mega-)	$10^{6}, 2^{20}$	M	나노 (nano-)	10^{-9}	n	zepto	10^{-21}	z
킬로 (kilo-)	$10^{3}, 2^{10}$	k, K	피코 (pico-)	10^{-12}	p	yocto	10^{-24}	y
헥토 (hecto-)	10^{2}	h				zetta	10^{21}	Z
데카 (deca-)	10^{1}	da, D				yotta	10^{24}	Y

주의 :

(1) $1 (= 10^0)$을 나타내는 별도의 접두어는 없다.
(2) 2의 지수로 나타낸 것은 data speed 를 나타낼 때 사용된다.
(3) 기호로 표시할 때 킬로(killo) 이상은 영어 대문자를 그 이하는 주로 영어 소문자를 사용한다.
(4) "킬로" 일 때는 10^3 은 소문자 k 를, 2^{10} 은 대문자 K를 사용한다.

3.6 수를 나타내는 접두사

1부터 20까지의 접두사와 실례

1	mono	11	hendeca
2	di(bi)	12	dodeca
3	tri	13	trideca
4	tetra	14	tetradeca
5	penta	15	pentadeca
6	hexa	16	hexadeca
7	hepta	17	heptadeca
8	octa	18	octadeca
9	ennea(nona)	19	nonadeca
10	deca	20	icosa

Chapter 3 수의 명칭

의미 (meaning)	접두사 (prefixes)	실례(examples)
$\frac{1}{2}$	semi-, demi-, hemi-	semicircle(반원), hemisphere(반구), demisemi(반의 반의, $\frac{1}{4}$ 의)
1	mono-, uni-	monomial(단항식), unique(유일한)
2	di-, bi-, du-	binomial(2항식), bicentennial(200년마다의), dipolar(쌍극의), dihedral(2면각)
3	tri-	triangle(삼각형), tricycle(3발자전거)
4	tetra-, quadri-	quadrant(4분면), tetragon(4각형), tetrahedron(4면체), quadric(2차의)★ quadruple(4곱절)
5	penta-, quinque-	pentagon(5각형), quinquennial(5년마다의), pentahedron(5면체), quintuple(5곱절)
6	hexa-, sex-	hexagon(6각형), hexahedron(6면체), sextuple(6곱절), sexangle(6각형)
7	hepta-, septa-	heptagon(7각형), septet(7중창), septagon(7각형), heptahedron(7면체)
8	octa-	octahedron(8면체)
9	nona-	nonagon(9각형), enneahedron(9면체)

● 199

원서읽기를 위한 수학용어사전

의미 (meaning)	접두사 (prefixes)	실례(examples)
10	deca-	decahedron(10면체), decagon(10각형), decade(10년간)
11	undeca-	undecagon(11각형)
12	dodeca-	dodecahedron(12면체), dodecagon(12각형)
13	trideca-	triskaidecagon(13각형)
14	tetradeca-	tetrakaidecagon(14각형)
15	pentadeca-	pentadecagon(15각형)
16	hexadeca-	hexadecimal(16진법)
20	icosa-	icosahedron(20면체)
small	micro-	microscope(현미경)
large	macro-	macrocosm(대우주)
little	mini-	minikin(작은 물건)
many	multi-, poly-	multiplex(복합적인), polynomial(다항식)

Chapter 3 수의 명칭

pentagram

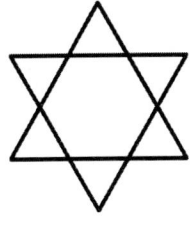
hexagram

3.7 진수 명칭(Base names)

Base 1	(unary)
Base 2	(binary)
Base 3	(ternary / trinary)
Base 4	(quaternary)
Base 5	(quinary / quinternary)
Base 6	(senary / heximal / hexary)
Base 7	(septenary / septuary)
Base 8	(octal / octonary / octonal / octimal)
Base 9	(nonary / novary / noval)
Base 10	(decimal / denary)
Base 11	(undecimal / undenary / unodecimal)
Base 12	(duodecimal / dozenal / duodenary)
Base 13	(tridecimal / tredecimal / triodecimal)
Base 14	(tetradecimal / quadrodecimal / quattuordecimal)

Base 15	(pentadecimal / quindecimal)
Base 16	(hexadecimal / sexadecimal / sedecimal)
Base 17	(septendecimal / heptadecimal)
Base 18	(octodecimal / decennoctal)
Base 19	(nonadecimal / novodecimal / decennoval)
Base 20	(vigesimal / bigesimal / bidecimal)
Base 21	(unovigesimal / unobigesimal)
Base 22	(duovigesimal)
Base 23	(triovigesimal)
Base 24	(quadrovigesimal / quadriovigesimal)
Base 26	(hexavigesimal / sexavigesimal)
Base 27	(septemvigesimal / heptovigesimal)
Base 30	(trigesimal / triogesimal)
Base 32	(duotrigesimal)
Base 36	(hexatridecimal / sexatrigesimal / hexatrigesimal)
Base 60	(sexagesimal)
Base 70	(septemgesimal / heptagesimal)
Base 80	(octagesimal / octogesimal)
Base 90	(nonagesimal / novagesimal)
Base 100	(centimal / centesimal)

Chapter 3 수의 명칭

진수		Greek	Latin
0			
1	unary	mono–	uni–
2	binary	di–	duo–, bi–
3	ternary	tri–	ter–, tre–,
4	quaternary	tetra–	quadri–, quadr–
5	quinary	penta–, pent–	quinqu–, quinque–
6	senary	hexa–, hex–	sexa–, sex–
7	septenary	hepta–, hept–	septua–
8	octal	octa–, oct–	octo–, oct–
9	novenary	ennea–	nona–
10	decimal	deca–, deka–	deci–
11	undecimal	hendeca–, hendeka–	undeca–
12	duodecimal		
13	tredecimal		
14	tetradecimal		
15	pentadecimal		
16	hexadecimal		
20	vigesimal		
24	quadrovigesimal		
60	sexagesimal		
100	centimal	hecto–, hect–	centi–
1000	millesimal		

마야 (20진법)	크기		발음
	1	1	hun
	20	20	kal
	20^2	400	bak
	20^3	8000	pic
	20^4	160000	calab
	20^5	3200000	kinchel
	20^6	64000000	alce

다항식	방정식
	1차방정식 (linear equation)
2차다항식 (quadratic polynomial)	2차방정식 (quadratic equation)
3차다항식 (cubic polynomial)	3차방정식 (cubic equation)
4차다항식 (quartic polynomial)	4차방정식 (quartic equation)
5차다항식 (quintic polynomial)	5차방정식 (quintic equation)

3.8 여러 나라 언어에서의 수사 (1부터 10까지)

	영어	독일어	프랑스어	스페인어	Greek	일본어
1	one	eins	un, une	uno, una	e s	いち (이찌)
2	two	zwei	deux	dos	d o	に (니)
3	three	drei	trois	tres	tre s	さん (산)
4	four	vier	quatre	cuatro	t ssares	よん (욘)
5	five	f nf	cinq	cinco	p nte	ご (고)
6	six	sechs	six	seis	hex	ろく (로꾸)
7	seven	sieben	sept	siete	hept	なな (나나)
8	eight	acht	huit	ocho	okt	はち (하찌)
9	nine	neun	neuf	nueve	enn a	きゅう (큐-)
10	ten	zehn	dix	diez	d ka	じゅう (쥬-)

Chapter 3 수의 명칭

	이탈리아어	러시아어	폴란드어	산스크리트어	고대그리스어
1	uno	один (odyn)	jeden(예덴)	eka	en
2	due	два (dva)	dwa(드바)	dva	duo
3	three	три (tri)	trzy(트씨)	tri	tri
4	quattro	четыре (chetyre)	cztery (츠테리)	catur	tetra
5	cinque	пять (piat)	pięć (피엥치)	panca	pente
6	sei	шесть (shest)	sześć (시에시치)	sas	hex
7	sette	семь (sem)	siedem (시에뎀)	sapta	hepta
8	otto	восемь (vosem)	osiem (오시엠)	asta	octo
9	nove	девять (deviat)	dziewięć (지에비엥치)	nava	ennea
10	dieci	десять (desiat)	dziesięć (지에시엥치)	daca	deca

	중국어	아랍어	Latin	포르투갈어	베트남어
1	一 yī (이)	١ (와헤드, 기니)	unus	um	một
2	二 r (얼)	٢ (이뜨닌)	duo	dois	hai
3	三 sān (싼)	٣ (탈라따)	tres	trs	ba
4	四 s (쓰)	٤ (알바하)	quatuor	quatro	bốn
5	五 wǔ (우)	٥ (깜싸)	quinque	cinco	năm
6	六 li (리우)	٦ (쎄따)	sex	seis	s u
7	七 qī (치)	٧ (싸바)	septem	sete	bảy
8	八 bā (빠)	٨ (다마니아)	octo	oito	t m
9	九 jiǔ (지우)	٩ (디싸)	novem	nove	ch n
10	十 sh (스)	١٠ (아샤하, 아시프)	decem	dez	mười

아라비아 숫자의 기원은 인도라고 하며, 아랍어 자체는 쓰고 읽을 때 오른쪽부터 왼쪽으로 나가지만, 숫자는 반대로 쓰고 읽는 것이 왼쪽부터 오른쪽이므로 주의해야 한다.

3.9 수를 포함하는 영어 단어

bicycle	2륜 자전거
bilingual	두 언어를 자유로이 구사하는
binomial	이항의
biquadratic	네제곱
bisect	이분하다
centipede	지네 (지네는 다리가 30-40개이다)
cubic [kjúːbik]	3차의
Decalogue [dékəlɔ̀ːg]	[성서] (모세의) 십계명(the Ten Commandments), 십계
decimal	십진법의
deuce	카드 2 (가장 낮은 카드)
dilemma	di + lemma, 딜레마 (참고) trilemma
dodecagon	12각형
duodenum [djdjuːádənəm]	십이지장 (길이가 12 인치)
enneagon [éniəgàn / -gən]	9각형, 9변형
fortnight	2주일, 14일
four score	4×20, 80

Chapter 3 수의 명칭

hecatomb	대학살, 신에게 바친 황소 100 마리의 제물
hectare	헥타르
heptadecagon	17각형
heptagon	7각형
hexameter [heksǽmitər]	[운율] 육보격(六步格), 육보격의 시, 6음보
hexapoda	6개의 다리를 가진 동물
Lent	사순절 (Ash Wednesday부터 Easter Eve까지의 40일)
millipede	노래기 (노래기는 다리가 200개 가량이다)
monologue	독백
monopoly	전매
myriapoda	다족류 (1만개의 다리를 뜻함)
nonagenarian	90세의 사람
octopus	문어
pentacle	별 모양의 5개의 뾰족한 끝
pentagon	오각형
pentagram	5각형 별 모양
Pentateuch [péntətjù:k]	[성서] 모세 5경(經)(구약성서의 첫 5편)
pentathlon	오종경기
Pentecost	오순절

quadrangle	사각형
quadratic [kwɑdrǽtik]	2차의
quadruped	네발짐승
quarter	1/4
quartic [kwɔːtik]	4차의
quintic [kwʌntik]	5차의
score	20
Septuagint	70인역 그리스어 성서
sextain	6행연
sextet	6중주
snake's eyes	두 개의 주사위를 던져 1의 눈이 2개 나오는 것
tetrapoda	4개의 다리를 가진 동물
three score and ten	인간 70평생
tripod	삼각대
triskaidekaphobia	13공포증, 13을 두려워 함 = paraskevidekatriaphobia = friggatriskaidekaphobia = triskaidekaphobia
undecagon	11각형
unicorn	유니콘

Chapter 3 수의 명칭

3.10 다각형 명칭

다각형(polygon) 명칭			
10 의 자리		1의 자리	
10 +	(예외) + decagon	+ 1	+ henagon
20 +	icosi +	+ 2	+ digon
30 +	triaconta +	+ 3	+ trigon
40 +	tetraconta +	+ 4	+ tetragon
50 +	pentaconta +	+ 5	+ pentagon
60 +	hexaconta +	+ 6	+ hexagon
70 +	heptaconta +	+ 7	+ heptagon
80 +	octaconta +	+ 8	+ octagon
90 +	enneaconta +	+ 9	+ enneagon
100 +	hecta +	+ 0	+ gon

　　다각형은 삼각형부터 생각할 수 있으며 1부터 10각형까지의 명칭은 약간 차이가 있을 수 있고, 11각형부터 19각형까지는 따로 기억하고 20각형은 철자를 주의해야 한다. 나머지는 십의 단위와 일의 단위를 결합하면 된다. 30부터 99까지는 다각형의 중간에 conta 가 들어가는 것이 재미있다. 예를 들어 65각형은

$$\text{hexaconta + pentagon = hexacontapentagon}$$

이 된다.

원서읽기를 위한 수학용어사전

3각형	trigon 또는 triangle	17각형	heptadecagon
4각형	tetragon 또는 quadrilateral	18각형	octakaidecagon
5각형	pentagon	19각형	enneadecagon
6각형	hexagon	20각형	icosagon★
7각형	heptagon 또는 septagon	30각형	triacontagon
8각형	octagon	40각형	tetracontagon
9각형	enneagon 또는 nonagon	50각형	pentacontagon
10각형	decagon	60각형	hexacontagon
11각형	hendecagon 또는 undecagon	70각형	heptacontagon
12각형	dodecagon	80각형	octacontagon
13각형	triskaidecagon	90각형	enneacontagon
14각형	tetrakaidecagon	100각형	hectagon
15각형	pentadecagon	n-gon	
16각형	hexakaidecagon		

Chapter 3 수의 명칭

3.11 수식 읽기와 주의사항

100	1백	hundred
1,000	1천	thousand
10,000	1만	ten thousand
100,000	1십만	one hundred thousand
1,000,000	1백만	million
10,000,000	1천만	ten million
100,000,000	1억	hundred million

주의 The word "and" does not appear in word names for whole numbers.

billion, million, thousand, hundred 다음에 and 를 넣지 않는다. 이때 and 를 넣으면 소수점으로 오해할 소지가 있다.

주의 dozen, hundred, thousand 는 수사 또는 수를 나타내는 어구와 함께 써도 복수형의 s를 붙이지 않는다.

주의 dozen 불특정 다수를 나타낼 때는 dozens 와 같이 s를 붙인다.

예 some dozens of people   수십 명의 사람들
some dozen of people   12명 (가량의) 사람들

46,625,314,732 는 "Forty-six billion, six hundred twenty-five million, three hundred fourteen thousand, seven hundred thirty-two" 로 읽는다.

2^{64} = 18,446,744,073,709,551,616 는 18 quintillion 446 quadrillion 744 trillion 73 billion 709 million 551 thousand 616 으로 읽는다.

Chapter 4 도량형 변환과 근삿값

Chapter 4 도량형 변환과 근삿값

영어로 된 수학책에는 m(미터), kg(킬로그램), l(리터) 대신에 feet, pound, gallon 을 주로 사용하는 것을 볼 수 있다. 따라서 기계적으로 단위를 환산하는 방법보다는 자연스럽게 영어로 된 수학책에서 나오는 문장을 살펴보는 것이 도움이 될 것 같다.

4.1 길이, 넓이, 부피, 무게의 환산

미터법-영국 도량형법 변환 (Metric Imperial Conversions)

➤ 길이(length)

1 inch = 2.54 cm

1 foot = 12 inches = 30.48 cm.

1 yard = 3 feet = 91.44 cm

1 rod = $5\frac{1}{2}$ yards = 502.92 cm

1 furlong = 40 rods = 20116.8 cm

1 mile = 8 furlong = 1609.344 m

1 yard is 36 inches.

1 meter is about 39.37 inches. (thirty-nine point three seven inches)

➢ 넓이(area)

1 square foot = 144 square inches

1 square yard = 9 square feet

1 acre = 4840 square yards

1 square mile = 640 acres

➢ 부피(volume)

1 tablespoon = 3 teaspoons

1 cup = 16 tablespoons

1 pint = 2 cups = 16 ounce(온스)

1 quart = 2 pints(파인트)

1 gallon = 4 quarts ≈ 3.7853 liter,

1 peck = 2 gallons

1 bushel = 4 pecks

1 quart = $\frac{1}{4}$ gallon = $\frac{1}{8}$ peck = 2 pints

1 gallon = 4 quarts = 8 pints = 2^7 ounce

1 peck ≈ 8.81 liters

1 peck = 2 gallons

1 pint(파인트) = $\frac{1}{2}$ quart = 16 ounce

Chapter 4 도량형 변환과 근삿값

1 ounce = $\frac{1}{16}$ pint

1 liter is about 1.06 quarts.

참고 위 결과를 보면 아래 앞의 단위를 2배하면 바로 뒤의 단위가 되는 것을 알 수 있다.

gill, chopin, pint, quart, pottle, gallon, peck, demibushel, firkin(=또는 bushel), kilderkin, barrel, hogshead, pipe, tun

이에 대하여 하나씩 알아보기로 한다.

● gill [dʒil] 질(액량의 단위; = 1/4 pint; ((미국) 0.118l, (영국) 0.142l)

● chopin 오래된 액량의 단위로 영국에서는 대략 quart 에 해당하며, 스코틀랜드에서는 1/2 pint 에 해당한다.

● pint [paint] 파인트(액량의 단위; = 1/2 quart, 4 gills; 간단히 pt.로 적는다.

● quart [kwɔːrt] 쿼트(액량인 경우는 1/4 gallon, 약 1.14l); 건량(乾量)(보리·콩 따위에서는 1/8 peck, 2 pints)

● pottle [pɑ́tl / pɔ́tl] 액량(液量)의 이름 (반 갤런); 반 갤런들이 컵

● gallon [gǽlən] 갤런(용량의 단위로 4 quarts; 간단히 gal., gall.로 적는다.

imperial gallon 영국 갤런(4.546l).

wine gallon 미국 갤런(3.7853l).
- peck 펙(영국에서는 9.092 리터; 미국에서는 8.81 리터).
- demibushel : 1/2 bushel
- firkin [fə́:rkin] (버터 따위를 담는) 조그만 나무통(8-9 갤런 들이); 용량의 단위(1배럴의 1/4).
- bushel1 [búʃəl] 부셸(약 36리터, 약 2말)
- kilderkin [kíldərkin] 통(16 또는 18갤런들이)
- barrel [bǽrəl] 1배럴(액량·건량의 단위: 영국에서는 36, 18 또는 9갤런; 미국에서는 31.5 갤런; 〖석유〗 42 미 갤런, 35 영 갤런).
- hogshead [-zhed] 큰통(영국 100-140 갤런들이; 미국 63-140 갤런들이); 액량(液量)의 단위(미국 63갤런; 영국 52.5갤런); 맥주·사이다 등의 단위(245.4리터; 영국 54갤런). hhd.로 생략함.
- pipe [paip] 큰 술[기름] 통; 그 큰 통의 용량 (미국) 126 gallons, (영국) 105 gallons.
- tun [tʌn] 큰 통, 큰 술통; (양조용의) 발효(醱酵)통; (술 따위의) 용량 단위(252갤런).

> 무게(weight)

1 grain = 0.0648 gram (밀 한 알의 무게)

1 ounce = $437\frac{1}{2}$ grains

1 pound = 16 ounces

1 hundredweight = 100 pounds

Chapter 4 도량형 변환과 근삿값

1 ton = 20 hundredweights

1 kilogram is about 2.2 pounds.

- 1 다스(dozen) = 12 개
- 12 다스 = 144개 = 1 그로스(gross)

4.2 근삿값 계산에 유용한 것들

- $e = 2.7\ 1828\ 1828\ 45\ 90\ 45\ \cdots$
- $e \approx \dfrac{878}{323} \approx 2.71826 \cdots$
- $e^3 \approx 20.0855 \approx 20$
- $\pi \approx \dfrac{22}{7},\ \dfrac{355}{113}$
- $2^{10} \approx 10^3$
- $2^7 \approx 5^3$
- $524{,}288 = 2^{19} \approx 3^{12} = 531{,}441$
- $\pi^2 \approx 10 \approx \dfrac{227}{23}$
- $\pi^4 \approx \dfrac{2143}{22}$
- $\pi^4 + \pi^5 \approx e^6$
- $e^\pi - \pi \approx 20$
- $\dfrac{\pi^9}{e^8} \approx 10$

- $100! \approx 9.33 \times 10^{157}$

- $(10{,}000)^{\frac{1}{9}} \approx \pi$

- $(1{,}000{,}000)^{\frac{1}{14}} \approx \pi$

- $\sqrt{2} \approx \dfrac{17}{12}$ 이다. 왜냐하면 $2 = \dfrac{288}{144} \approx \dfrac{289}{144} = \left(\dfrac{17}{12}\right)^2$.

- $\sqrt{2} \approx \dfrac{239}{169}$ 이다. 왜냐하면 $2 = \dfrac{57122}{28561} \approx \dfrac{57121}{28561}$.

- $\sqrt{2} \approx \dfrac{3363}{2378}$

- $365 \times 274 = 100{,}010 \approx 10^5$

- $111{,}111{,}111 \times 111{,}111{,}111 = 12{,}345{,}678{,}987{,}654{,}321$

- 1평은 $\left(\dfrac{20}{11}\right)^2 m^2$ 로 $3.305785\ m^2$ 이다.

- **황금비** Mathematically speaking, two quantities are in the Golden Ratio if their ratio is the same as the ratio of their sum to the larger of the two quantities.

: 황금비는 본래 $x^2 - x - 1 = 0$ 의 양의 근인 $\dfrac{1 + \sqrt{5}}{2}$ 로 주어지며 수가 아니라 비례관계이다.

☞ 5 miles $\approx$ 8 km 이므로, 1 mile $\approx$ 1.6 km 로, 1 mile 과 1 km 의 비는 대략 황금비와 같다.

Chapter 4 도량형 변환과 근삿값

4.3 단위 변환

- 1인치 = 2.54cm
- 1ft = 30.48cm = 0.3048m
- 1m ≈ 39.17인치 ≈ 3.281ft
- 1 m/h ≈ 0.4470 m/s
- 1 마일 = 5280 피트
- 1해리 = 1,852 m
- 1 km = 약 0.54 해리 (근사적으로 2 km 가 1해리이다.)
- 지구 (태양주위를 도는) 공전속도는 초속 18.5 마일
- 금성 (태양주위를 도는) 공전속도는 초속 22마일
- 태양의 적도 반지름은 695,553 km 이다.
- 1광년 ≈ $9.46053 \times 10^{12}\ km$ ≈ $9.5조 km$ ≈ 5.9×10^{12} 마일
- 지구와 태양과의 평균거리(1 천문단위 : 1 AU)는 대략 149,597,870 km 이다.

☞ 1 AU ≈ 1.5×10^8 km ≈ 93,000,000 마일 = 9.3×10^7 mile

☞ The distance from the earth to the sun (= 1 AU) is about 93,000,000 miles.

- 지구 적도반지름 ≈ 6.3781×10^8 cm (약 6378 km) ≈ 3960 마일
- 마하(Mach number)는 소리의 속도가 340 m/sec 이므로 "1 마하(Mach) = 1,224 km/h"가 된다. 마하 1 보다 큰 경우 초음속이라 한다.
- 지구의 자전속도는 시속 1,667 km 정도 (= 40,000 km /

24 시)이며 이것은 마하 1.36 이다.
- 지구의 적도(위도 0도)에서 경도 1도에 대한 호의 길이는 111.306 km이며, 위도 45도에서는 78.837 km 이고, 위도 89도에서는 1.949 km 이다.
- 달의 질량은 7.349×10^{22} kg, 지구의 질량은 5.976×10^{24} kg이고, 달의 반지름은 1737 km, 지구의 반지름은 6378 km 이다.
- 지구는 태양주위를 시속 107,245 km 로 움직인다.
- 야구장의 내야는 한 변의 길이가 90 ft 인 사각형이다.
- 에펠탑(Eiffel Tower)의 높이는 1,063 ft (약 324 m)이다.
- 달에서의 중력가속도는 1.6 m/sec이다.
- 에베레스트 산의 높이 : 29,029 feet
- 지구에서 북극성(North Star)까지 거리는 약 47 광년(light year)이다.
- 센타우루스자리 프록시마(Proxima Centauri)는 태양에서 가장 가까운 별이며, 센타우루스자리의 알파별까지는 (지구로부터) 약 4.3 광년 거리에 있다.
- 1 년 = 12 달 ≈ 365.25 일 = 8,766 시간 = 525,960 분 = 31,557,600 초 ≈ 3×10^7 초 ≈ $\pi \times 10^7$ 초
- 빛의 속도 299,792 km/s ≈ 186,282 mile/sec ≈ 1.8×10^5 mile/sec
- 쌍곡선 함수로 유명한 미국 미주리주 St. Louis 에 있는 Gateway Arch는 높이가 630 ft 이다. 이것은

Chapter 4 도량형 변환과 근삿값

$$y = -127.7\,ft \cdot \cosh\left(\frac{x}{127.7}\,ft\right) + 757.7\,ft$$

로 나타난다. 여기서 $\cosh x = \dfrac{e^x + e^{-x}}{2}$ 이다.

- $y = \sin x$ (x는 라디안)의 $[0, \pi]$ 부분의 x축과 이루는 부분의 넓이는 2이다.
- (섭씨와 화씨)

☞ To convert from degrees Celsius to degrees Fahrenheit use $F = \dfrac{9}{5}C + 32$.

☞ To convert from degrees Fahrenheit to degrees Celsius use $C = \dfrac{F-32}{9} \times 5$.

주의 $-40\,°C = -40\,°F$, 즉 $-40\,°$에서 섭씨온도와 화씨온도가 수치적으로 같다.

water boils : $100\,°C = 212\,°F$
normal body temperature : $37\,°C = 98.6\,°F$
water freezes : $0\,°C = 32\,°F$

섭씨(도)	−10	0	10	20	30	40	100
화씨(도)	14	32	50	68	86	104	212

- 철길의 레일 간의 궤간거리 → 우리나라는 표준궤 1.435 m 를 사용한다.

1.435 m = 4피트 8.5인치 = 56.5인치

- The phrase "rule of thumb (엄지손가락의 법칙, 주먹구구)" is derived from and old English law which stated that you couldn't beat your wife with anything wider than your thumb.

- 현재 사용되는 미국의 지폐는 $1, $2, $5, $10, $20, $50, $100 의 7종으로 그 크기는 모두 같고, 가로 $6\frac{1}{4}$ 인치, 세로 $2\frac{5}{8}$ 인치 (6.25인치× 2.625인치)이다.

☞ A dollar bill weighs about 1 gram. (1달러 지폐의 무게는 약 1 그램이다.)

- 현재 미국에서 통용되는 동전은 penny (1 cent), nickel (5 cent), dime (10 cent), quarter (25 cent), half dollar (50 cent) 5종류이다. [물론 dollar (100 cent) 동전도 발행되고 있지만 잘 사용되지 않는다.] **(309 쪽 참조)**

☞ 미국 자판기에는 nickel, dime, quarter 이 사용된다.

☞ A nickel weighs about 5 grams. (5센트 동전의 무게는 약 5 그램이다.)

- There are 293 ways to make change for a dollar.
  : 1달러 지폐를 이들 5종류의 동전으로 교환하는 방법의 수는 293 가지가 된다.

- square 는 대수에서는 "제곱"을, 기하에서는 "정사각형"을

Chapter 4 도량형 변환과 근삿값

의미한다.
- tangent 는 수학에서 "탄젠트"라는 의미와 "접선"이라는 의미가 있다.
- secant 는 수학에서 "시컨트"라는 의미와 "할선"이라는 의미가 있다.
- 임의로 선택한 두 정수가 서로소일 확률은 $\dfrac{6}{\pi^2} \approx 0.6079$ 이다.
- Light travels 186,000 miles in 1 second.
: 빛은 초속 186,000 마일이다.
- Sound travels 1100 feet in 1 second.
: 소리는 초속 1100 피트이다.
- 마라톤 거리 42,195 km = 26 마일 385 야드, 약 26.2 mile
- A판 용지와 B판 용지

: 독일 산업 표준 (DIN : Deutsche Industrie Normung; German Industrial Standard)에 의해 1922년에 A 크기의 종이 크기를 도입한 표준이 되었으며 1975년에 국제 표준 ISO 216에 채택되었다. (A판과 B판 모두 세로와 가로의 비가 $\sqrt{2}$: 1 인 닮음 직사각형이다.)

☞ A_0 1189㎜ × 841㎜ 로 넓이는 1 m^2 이다.
☞ B_0 1456㎜ × 1030㎜ 로 넓이는 1.5 m^2 이다.

- 우리나라에서 사용하는 단위
  (1) 한약의 경우 : 1 제(劑) = 20 첩(貼)

    주의 이때 '재'가 아님을 주의해야 한다.

(2) 담배의 경우 : 1 갑(匣) = 20 개비

주의 : 이때 '개피' 대신 '개비'이다.

20 개비는 하루에 다 피우기 적당한 양이라고 한다.

(3) (말린) 오징어의 경우 : 1 축 = 20 마리
(4) 조기와 청어 등의 경우 : 1 두름 = 20마리
(5) 근(斤) : 저울로 다는 무게의 단위, (보통 고기류에서) 600g 또는 (야채에서는) 375g 이다.

> **참고** 대만에서는 1근이 600 克(gram)이고 중국(본토)에서는 500 克(gram)이다.

(6) 관(寬) : 무게의 단위로 1관은 3.75Kg이다.

Chapter 5 수학 수식 읽기

Chapter 5 수학 수식 읽기
(수식과 기호 읽기)

이 장에서는 수식과 그 읽기에 대해 알아보기로 한다.

$+$	plus, add, positive value
$-$	minus, subtract, negative value
$\pm$	plus or minus, add or subtract
$\mp$	minus or plus
$\times$, $\cdot$	times, multiplied by
$\div$, $/$, $-$	divided by
$=$	equals, is equal to
$\neq$	is not equal to
$\equiv$	is identical with
$\approx$, $\fallingdotseq$	is approximately equal to
$\sim$	is equivalent, is similar to
$>$	is greater than
$\gg$	is much greater than
$<$	is less than
$\ll$	is much less than
$\not>$	is not greater than
$\not<$	is not less than
$\geq$, $\geqq$	is greater than or equal to

원서읽기를 위한 수학용어사전

$\leq$, $\leqq$	is less than or equal to
$\propto$	varies directly as, is directly proportional to
$\rightarrow$	approaches as a limit
( )	parentheses　　(단수) parenthesis
[ ]	bracket(s)
{ }	brace(s)
—	vinculum　　(예) $\overline{a+b}$
! , $\lfloor$	factorial of
∞	infinity
$4,000,000,000$	four billion
$6,324,859$	six million, three hundred twenty-four thousand, eight hundred fifty-nine
$2+3=5$	Two and [plus] three are [is, make(s), equal(s)] five.
$9-6=3$	① Six from nine leaves three. ② Nine minus six equals three.
$2 \times 4 = 8$	① Two times four is eight. ② Two multiplied by four is [equals] eight.

주의 $a \times b$ 에서 a 가 분수 또는 소수일 때는 (b 는 정수) "a of b" 로 읽고 b 가 분수 또는 소수일 때는 (a 는 정수) "a times b" 로 읽는다.

$8 \div 4 = 2$	① Four into eight goes twice. ② Eight divided by four makes two.
$a = b$	① a equals b ② a is identical with b
$3^2 = 9$	The square of three is nine.
$3^3 = 27$	① The cube of three is twenty-seven. ② Three cubed is twenty-seven.
$a = b^3$	a is b cube.

Chapter 5 수학 수식 읽기

$1 \times 0 = 0$	One multiplied by nought is nought.
a^2	a square, a squared
a^3	a cube, a cubed
a^4	a [raised] to the fourth [power]
a^{-1}	a to the minus one
$f(x)$	a function of x
$\angle$	angle　　　　(복수) $\angle$ s
$\perp$	is perpendicular to　(복수) $\perp$ s
$\parallel$	is parallel to　(복수) $\parallel$ s
$\triangle$	triangle　　　(복수) $\triangle$ s
▭	rectangle
□	square
▱	parallelogram
○	circle　　　　(복수) ⓢ
$\cong$	is congruent to
$\sim$	is similar to
$\therefore$	therefore, hence
$\because$	since, because
$\overset{\frown}{GH}$	the arc between points G and H
$\triangle y$	an increment of y

dx	differential of x
$\sum$	summation operator
$\prod_1^n$	product of n terms
$7:4$	the ratio of seven to four
$3:6=4:8$	이때 : 는 is to 또는 the ratio of 로 읽는다. ① Three <u>is to</u> six as four <u>is to</u> eight. ② <u>The ratio of</u> three to six equals <u>the ratio of</u> four to eight. ③ three divided by six equals four divided by eight.
$'$	prime
$''$	double prime
$'''$	triple prime
$\int f(x)\,dx$	the integral of $f(x)$ with respect to x
$\left(5 + 2\dfrac{1}{6} - 3.33 \times 2\right) \div 7\dfrac{1}{3}$	five plus two and one sixth minus three decimal three three multiplied by two, all divided by seven and a third.
$\sqrt{4}=2$	The square root of four is two.
$\sqrt{a}$	square root of a ("root of" 대신에 "radical" 이라고도 한다.)
$\sqrt[3]{b}$	cube root of b
$\sqrt[m]{n}$	m root of n
$\sqrt[3]{1000}$	the cube root of a thousand
0.01	nought point nought one
0.12	nought point one two

Chapter 5 수학 수식 읽기

2.1	① two point one ② two and one-tenth
$\dfrac{1}{2}$	① a half ② one-half
$\dfrac{1}{3}$	① a third ② one-third (주의) 미국에서는 분수에 하이픈(-)을 하지 않는다.
$\dfrac{1}{4}$	① a quarter ② one-fourth
$\dfrac{3}{7}$	three-sevenths
$8\dfrac{2}{5}$	eight and two-fifths
$\dfrac{210}{365}$	two hundred (and) ten over three hundred (and) sixty-five
$\dfrac{c}{a+b}$	c over a plus b
$\dfrac{a}{b}$	a upon b
$5^2 = 25$	Five squared is twenty-five.
$\sqrt{36} = 6$	The square root of thirty-six is six.
$a^2 - b^2 = (a+b)(a-b)$	a square minus b square equals parenthesis a plus b parenthesis times parenthesis a minus b parenthesis.
$(a+b)^2 = a^2 + 2ab + b^2$	The square of the sum of a plus b equals a square plus two ab plus b square.
$\|a\|$	absolute value of a
%	percent
a^n	a to the n-th (power)

분수와 소수 읽기

주의 분수를 읽을 때는 분자(numerator)는 기수(cardinal numbers)로 읽고, 분모(denominator)는 서수(ordinal numbers)로 읽는다. 단 이때 분자가 2 이상이면 분모를 복수형으로 한다.

$\dfrac{1}{7}$	a [또는 one] seventh
$\dfrac{2}{7}$	two sevenths
$\dfrac{53}{100}$	fifty-three hundredths
11.7	eleven point seven
28 %	twenty-eight per cent
$1\dfrac{2}{3}$	one and two thirds
$\dfrac{4}{9}$	four ninths
6.25	six point two five
0.01	point zero one
$4.\overline{3}$	four point three repeating (repeating 대신에 recurring 또는 circulating 을 사용하기도 한다.)
$3.1\overline{45}$	three point one, forty five repeating (또는 recurring 또는 circulating)
$\dfrac{1}{4}$	a quarter
$\dfrac{3}{4}$	three quarters

Chapter 5 수학 수식 읽기

$\dfrac{1}{10}$	tenths	
$\dfrac{1}{100}$	hundredths	
$\dfrac{1}{1,000}$	thousandths	
$\dfrac{1}{10,000}$	ten thousandths	
$\dfrac{1}{100,000}$	hundred thousandths	
$\dfrac{123}{456}$	123 over 456 또는 123 by 456	
397.685	three hundred ninety-seven <u>and</u> six hundred eighty-five <u>thousandths</u>	
413.87	four hundred thirteen and eighty-seven <u>hundredths</u>	
175	one hundred (and) seventy-five	
6,867	six thousand (and) eight hundred sixty-seven	
35,943	thirty-five thousand nine hundred and forty three	

● 수(number)와 숫자(numeral)의 차이

　　A number is an idea or a concept. A numeral is the symbol or representation for a number.

　: 수(number)는 눈으로 볼 수도 없고 손으로 만질 수도 없다. 그렇게 때문에 눈에 보이지 않는 수를 볼 수 있게 해주는 숫자(numeral)가 있어야 한다. 숫자(numeral)는 수(number)를 표현하기 위해 사용하는 기호로 인도-아라비아 숫자, 로마숫자 등이 있다. 참고로 아라비아 숫자는 편리함에도 불구하고 쉽게 다른

숫자로 고칠 수 있거나 잘못 읽혀질 수가 있다. 따라서 금융거래 특히 수표 등에 금액을 적을 때 위변조 방지를 위한 장치를 두고 있다.

☞ 수학에서 or 를 번역할 때
"또는" 과 "즉"이 있다. "또는"은 둘 중 하나에 속하는 경우로 두 집합의 합집합(union)일 때를 말하고, "즉"은 둘이 동치인 경우이다. 예를 들어
$2x + 3 = 11$ or $2x = 8$ or $x = 4$
는 $2x + 3 = 11$ 또는 $2x = 8$ 또는 $x = 4$ 보다는
$2x + 3 = 11$ 즉 $2x = 8$ 즉 $x = 4$
로 번역하는 것이 옳다.

☞ 수학용어로 "1부터 10까지"는 우리나라에서는 경곗값을 포함하지 않는다. 그러나 영어에서 "between 1 and 10"은 일반적으로 1과 10을 포함한다. 즉 영어로는 "between 1 and 10 (inclusive)"의 뜻이다. 따라서 경계를 포함하지 않는 "1부터 10사이"로는 "between 1 and 10 (exclusive)"를 이용한다.

☞ more than once 는 '2번 이상'으로 번역해야 좋을 듯하다.

☞ 더하기, 빼기, 곱하기, 나누기, 지수, 괄호 등이 섞인 혼합계산일 경우 계산 순서는 괄호, 지수, 곱하기와 나누기, 더

하기와 빼기 순으로 하기로 약속한다.

대부분 미국의 초등학교 교실 뒷벽에 적혀있는 PEMDAS 는 Please Excuse My Dear Aunt Sally.

로 기억하며 여기서 P는 parentheses(괄호), E는 exponents (지수), M은 multiplication(곱셈), D는 division(나눗셈), A 는 addition(덧셈), S는 subtraction(뺄셈)으로 연산의 우선 순위를 나타내고 있다.

● The order of operations agreement
(1) Do all operations inside parentheses (= grouping symbols).
(2) Simplify any number expressions containing exponents.
(3) Do multiplication and division as they occur from left to right.
(4) Do addition and subtraction as they occur from left to right.

☞ 로마숫자

I(1), V(5), X(10), L(50), C(100), D(500), M(1,000)이 주로 쓰이며 I, II, III, IV, V, VI, VII, VIII, IX, X, XI, XII 는 아직도 일부 벽시계에서 볼 수 있지만 나머지 큰 숫자들을 읽기가 불편하다. 특히 IV = 4, VI = 6, IX = 9, XI = 11, XC = 90, CX = 110 등 주의가 필요하다. 로마 숫자는 아라비아 숫자가 위변조에 취약하기 때문에 서양에서 예전에 많

이 사용하였으며 지금도 그 흔적이 많이 남아있는 수 표시 방법 중 하나이다.

I	V	X	L	C	D	M
1	5	10	50	100	500	1000

I	II	III	IV	V	VI	VII	VIII	IX	X
1	2	3	4	5	6	7	8	9	10

XI	XII	XIII	XIV	XV	XVI	XVII	XVIII	XIX	XX
11	12	13	14	15	16	17	18	19	20

로마 숫자는 5, 10, 50, 100, ⋯ 을 기준으로 덧셈과 뺄셈 방법에 따라 이루어졌다.

아라비아 숫자	구성		로마 숫자
4	(5−1)	−I+V	IV
9	(10−1)	−I+X	IX
40	(50−10)	−X+L	XL
90	(100−10)	−X+C	XC
400	(500−100)	−C+D	CD
900	(1000−100)	−C+M	CM

아라비아 숫자	구성		로마 숫자
6	(5+1)	V+I	VI
11	(10+1)	X+I	XI
60	(50+10)	L+X	LX
110	(100+10)	C+X	CX
600	(500+100)	D+C	DC
1100	(1000+100)	M+C	MC

Chapter 5 수학 수식 읽기

로마숫자

23	XXIII	1300	MCCC
24	XXIV	1400	MCD
27	XXVII	1900	MCM
33	XXXIII	2800	MMDCCC
34	XXXIV	3500	MMMD
39	XXXIX	4300	MMMMCCC
44	XLIV	5400	MMMMCD
49	XLIX	6600	MMMMMMDC
52	LII	7400	MMMMMMMCD
59	LIX	8200	MMMMMMMMCC
69	LXIX	9000	MMMMMMMMM
78	LXXVIII	9300	MMMMMMMMMCCC
84	LXXXIV	9800	MMMMMMMMMDCCC
93	XCIII	10000	MMMMMMMMMM
99	XCIX	10600	MMMMMMMMMMDC
104	CIV	11200	MMMMMMMMMMMCC
114	CXIV	11600	MMMMMMMMMMMDC
200	CC	11700	MMMMMMMMMMMDCC
400	CD	12000	MMMMMMMMMMMM
800	DCCC		
900	CM		

Chapter 6 영어에서 나타나는 수학용어

이 장에서는 영어로 수학용어가 일상생활 속에서 나타나는 표현들에 대하여 정리하였다. 일부는 짐작하여 뜻을 헤아릴 수 있지만, 대부분은 바로 뜻을 파악하기 어려운 것들이다.

▷ square meal : 실속 있는 식사, 푸짐한 식사
▷ call it square : 해결된 것으로 하다.
▷ square the circle : 원의 넓이와 같은 정사각형을 만들다.
    (3대 작도 불능문제와 같은) 가능하지 않은 일을 시도하다.
▷ Back to square one.
 : 다시 시작하다, 원점으로 돌아오다, back to the start
▷ square deal. : 공정한 거래
▷ Don't be square. = Don't be so old fashioned.
    (cf) square 는 "고지식하고 변통 없는 사람"이라는 뜻
▷ Two's company; three's a crowd.
 : 셋은 많고 둘이면 족하다.
▷ Put two and two together. : (추론하여) 옳은 결론을 내린다.

Chapter 6 영어에서 나타나는 수학용어

▷ in two twos : 즉시, 곧 = at once
▷ Two heads are better than one.
: 혼자의 생각보다 여러 사람의 꾀가 낫다.
▷ Two of a trade seldom (or never) agree.
: 같은 장사끼리는 의가 좋지 못하다.
▷ two shakes of a lamb's tail
: (양 꼬리가) 2번 흔들기 전에, 금방
　　(예) I will be back in two shakes of a lamb's tail.
▷ It takes two to tango.
: 손뼉도 마주쳐야 소리가 난다.
　탱고 춤을 추려면 두 사람이 필요하다.
▷ A bird in the hand is worth two in the bush.
: 덤불속의 두 마리 새보다 내 손안에 한 마리의 새가 더 낫다.
▷ Birds of a feather flock together.
: 깃털이 같은 새는 모인다. 가재는 게편
▷ A penny for your thoughts?
　= I would like to know what you are thinking.
: 멍하니 무슨 생각을 그렇게 하니?

> **참고** penny는 영국의 화폐 단위로 개수를 말할 때는 pennies를 쓰고, 금액을 말할 때는 pence를 쓴다. 2부터 11까지는 한 단어로 twopence, threepence, … , elevenpence 와 같이 한 단어로 쓰고, 나머지의 경우는 두 단어로 쓰거나 가운데 하이픈을 넣는다.

▷ A penny saved is a penny earned. : 돈을 절약하면 그만큼 번 것이다.

▷ Second banana. : (코미디 등에서) 조역

▷ second hand : (시계의) 초침

▷ second cousin : 육촌

▷ secondhand : 전해들은, 중고의

▷ Three cheers. :

　Three cheers for ~ : ~을 위해 만세삼창

▷ the three R's

: (기초 교육의) 읽기·쓰기·셈(reading, writing, arithmetic).

▷ the rule of three : 복비례

▷ the Three in One : 삼위일체 = the Trinity

▷ A third wheel. : 쓸모없는 사람, 무용지물

▷ five-by-five : 키가 작고 통통한

▷ Give me five. : 악수하자.

▷ take five : (5분간) 잠깐 쉬자.

▷ five-and-dime : 싸구려 백화점 = five-and-ten

▷ five-by-five : 키가 작고 똥똥한

▷ a five cornered square : 꽉 막힌 사람

▷ accipiter quinqueceps : 다섯 가닥 얼굴붕대, 오두붕대

▷ sixth sense. : 제6감, 직감

▷ at (또는 to) sixes and sevens : 난잡하게, 혼란스러운

▷ hit (someone 또는 something) for six : 대성공을 거두다.

▷ six of one and half-a-dozen of the other : 오십보백보

Chapter 6 영어에서 나타나는 수학용어

▷ A stitch in time saves nine.
: 적당한 때 한번 수고하면 나중에 아홉의 수고를 덜 수 있다.
▷ nine times out of ten : 십중팔구, 대개
▷ A cat has nine lives. : 고양이는 쉽게 죽지 않는다.
▷ up to the nines : 완전하게, 더할 나위 없이
▷ nine-to-five : 월급쟁이, 9시 출근 5시 퇴근
▷ nine days' wonder
: 잠시 큰 이야기 거리가 되었다가 잊어지게 되는 사건[일]
▷ on cloud nine : very happy
▷ Ten-Four : 알았음, 납득
▷ Point of no return. : 너무 깊숙이 들어가 나오기 어려운 상황
▷ Murder one. : 일급살인(=first-degree murder)
▷ as simple as one, two, three : 아주 간단한
  (예) as easy as ABC : 아주 쉬운
▷ divide and conquer. : 분단공략
▷ Multiply like rabbits. : 토끼와 같이 번식하다.
  (토끼는 번식이 빠르고 증가하는 속도가 대단하기 때문에)
▷ Another day, another dollar. : 매일 매일이 같은
  = everything is ordinary, same routine, 그럭저럭 사는
▷ An ounce of prevention is worth a pound of cure.
: 병이 생겨서 심각해지기 전에 미리 알아내서 예방하는 것이 병들어 치료 받는 것 보다 더 중요하다. 1온스의 예방은 1파운드의 치료만큼의 가치를 지닌다.
▷ a love triangle : 삼각관계

▷ the eternal triangle : (남녀의) 3각 관계

▷ a right triangle : 직각삼각형

▷ an acute triangle : 예각삼각형

▷ an obtuse triangle : 둔각삼각형

▷ Give him an inch, and he'll take a mile.

: 손목 주면 팔 달랜다. 조금 친절히 해주면 기어오른다.

▷ One-way mirror.

: 나는 상대를 볼 수 있는데 상대가 나를 볼 수 없는 경우에 one way mirror라 한다. (cf) two way mirror는 양 쪽 방향에서 다 볼 수 있는 것을 말한다.

▷ Not a second too soon.

: something happened when it was almost too late; just in time. (cf) moment 보다는 minute, minute 보다는 second 로 갈수록 짧은 시간을 나타낸다.

▷ forty winks. = short sleep ; snooze say, 잠깐 눈을 붙이다.

▷ first class : 최고급, 일류

▷ Half a loaf is better than none.

: 빵 반개라도 없는 것보다는 낫다.

▷ in the top ten : 상위 10개 안에

▷ looking out for number one : 자기 자신[이익]만을 생각하다.

▷ twenty-three skidoo

: 떠나다, 달아나다, let's get out of here

▷ Going (round) in circles.

: 같은 곳을 빙빙 맴돌고 있다.

Chapter 6 영어에서 나타나는 수학용어

결정을 내리지 못하고 허송세월하다.

▷ complete 180. : 정반대

 (cf) 180 에 degree 를 붙이지 않아도 180 도 태도를 바꾼다는 의미가 있음.

▷ One-way ticket.

: 편도 승차권, 피할 길을 주지 않는 확실한 방법

▷ one for the road = have a drink before leaving

▷ 수학자 J. H. Conway 식 표현

☞ one and only one : Exactly one, 오직 한번 = onee

☞ two and only two = twoo

☞ three and only three = threee

참고 J. H. Conway 식 표현이란 우리가 if and only if 를 iff 로 적는 것에 착안한 표현이다. 내가 사과를 두 개를 가지고 있다면 내가 사과를 하나를 가지고 있는 경우도 될 수 있다. 그러나 "I have two and only two apples."라고하면 내가 정확하게 사과를 두 개를 가지고 있는 경우이므로 사과를 하나만 가지고 있는 경우는 해당이 안 된다.

▷ Feel ten feet tall. = To be proud

▷ hundreds and thousands

: (과자 장식으로 뿌리는) 굵은 설탕

▷ like a thousand bricks, like a hundred bricks

: 맹렬한 기세로

▷ He [she] always gives 110 %.

: 그 [그녀]는 항상 최선을 다한다.

☞ give 110 % : make the maximum possible effort

I give 110 % for any task (that) I am given.

: 나는 주어진 어떤 일에도 최선을 다합니다.

▷ (at) the eleventh hour

: 마지막으로, 아슬아슬하게 시간 안에 마지막으로, 거의 늦을 뻔한

▷ fair and square : 공명정대한, 공명정대하게

▷ a ten-gallon hat

: A ten gallon hat is often thought to be large enough to hold ten gallons of water. This is not true (unless you have an exceptionally large head).

▷ third rate : 3류의, 열등한

▷ one in a million. : 백만명 가운데 하나, 아주 드문 사람[것]

▷ look like a million : 근사해 보이다.

▷ millipede, millepede : 노래기, 다리가 많은 동물

▷ one-way street : 일방통행로

▷ one way ticket : 편도승차권(미국) ― (영국) single ticket

▷ die a thousand deaths : 죽을 고생을 하다, 몹시 불안해하다

▷ 101 : ~개론, 입문, (대학의) 기초 과정의, 입문의,

(예) Architecture 101   건축학 개론

▷ play second fiddle

Chapter 6 영어에서 나타나는 수학용어

: 보조역을 담당하다, 들러리 (노릇)을 하다.
▷ I don't want to play second fiddle to you any more.
: 이제 더 이상 네 들러리를 서고 싶지 않다.
▷ I was told by the director to play second fiddle.
: 단역을 맡으라고 내게 감독이 말했다.
▷ Don't count your chickens before they hatch [are hatched].
: 김치 국부터 마신다. 병아리가 부화하기 전에 수를 세지 마라.
▷ seventh heaven : 최고의 행복, 하늘나라
▷ in seventh heaven
: 최고로 기쁜(extremely happy), 기분이 환상적인
▷ on cloud nine : 황홀경을 느낄 정도로 기분이 좋은
▷ ten to one : 십중팔구, 대개, something very likely
▷ the upper ten : 상류계급 = the upper thousand
▷ count up : 합계하다
▷ count down : 수를 거꾸로 읽다, 초읽기를 하다.
▷ count to ten : 마음을 진정시키기 위해 10 까지 세다.
▷ all in one piece : safely, 안전하게
▷ all in all : 무엇보다 소중한 것
▷ one and all : 모두
▷ One tree doesn't make a forest.
: 더 많은 증거를 가지기 전까지는 기다려야 한다.
▷ Once bit, twice shy.
: 자라보고 놀란 가슴 솥뚜껑 보고 놀란다.
: Once hurt, one is doubly cautious in the future.

▷ once in a blue moon : 매우 드물게 = seldom

▷ Two wrongs don't make a right.

: '악을 악으로 대한다고 선해지는 것이 아니다'

'남이 자기에게 잘못한다고 해서 똑같이 그 사람에게 잘못하면 그 결과가 좋게 나올 수는 없다.' 즉, '악을 악으로 대한다고해서 그 결과가 선이 되지는 않는다.'

▷ Kill two birds with one stone.

: 일석이조, 두 가지 일을 한 번에 해내다.

solve two problems with one move

▷ two cents' worth = unsolicited opinion (자발적인 의견)

= An individual's opinion

☞ If you want your opinion to be heard, then add your two cents worth in front of others.

: 당신의 의견을 알리고 싶다면 다른 사람들 앞에서 그 의견을 주장하여라.

☞ Give us your two cents. : 당신의 의견을 말하라.

☞ put in one's two cents = say your opinion

(예) I put in my two cents at the meeting.

▷ Stand on your own two feet.

: 자립하다. to be able to provide all of the things you need for living without help from anyone else

▷ Half a loaf is better than none.

: 조금이라도 있는 것이 없는 것 보다 낫다.

▷ A picture is worth a thousand words. (Napoleon Bonaparte)

Chapter 6 영어에서 나타나는 수학용어

: 나폴레옹이 말했다는 "Un bon croquis vaut mieux qu'un long discours." 로 "천 마디 말을 듣는 것보다 한 번 보는 게 더 낫다".

▷ second string = group of substitute players
: 대안, 조연 (동의어는 second team) (반대어는 first string)

▷ dressed to the nines
: 특별한 상황에서 아주 맵시 있는 옷을 입었다.

▷ to the nines = to perfection 또는 to the highest degree

▷ one for the book : 기록해 둘 만한 것, 굉장한 일, something remarkable; an amazing thing, case, etc
  (예) That storm was really one for the book.

▷ one-two punch : 좌우연타 펀치

▷ Three strikes and you're out. : 삼진 아웃제

▷ on cloud 9 : 의기양양하거나 매우 행복한 상태
  (미국기상청에서 사용한 용어)

▷ a catch-22 : 모순되는 규칙, 진퇴양난의 상태

▷ my better half : 배우자

▷ one-horse town : 매우 작은 도시, 작고 발전이 늦은 도시

▷ second childhood.
: 지적 기능이 감소하여 어린애기와 유사한 돌봄이 필요로 하는 어른을 나타냄.

▷ a hot number : 음악에서 매우 흥분되는 부분, 매력적인 여자

▷ one-track mind : 편협한 마음

▷ The odds are a million to one. : 승산은 백만 대 1이다.

● 245

▷ Fool me once, shame on you; fool me twice, shame on me.
: 사기를 한번 당하는 것은 당신의 잘못이 아니다. 그러나 똑같은 사기를 두 번 당하는 것은 당신의 책임이다.

▷ first string : 일급 선수

▷ second string : 대안, 조연

▷ third string
: 상태나 질이 낮은, first string 이나 second string 보다 낮은

▷ as easy as one, two, three = a piece of cake : 아주 쉬운

▷ one-night stand
: 하룻밤 흥행 (옛날 순회극단이 인기를 끌던 시절 작은 마을에선 하룻밤만 공연을 하고 떠나던 관행에서 유래)

▷ put two and two together and make five
: 특별한 상황을 잘못 이해하다.
 (cf) just like adding two and two

▷ batting a thousand
: (대성공을 거두다) (타율 1.000 즉 10할 타율을 말한다.)

▷ Fifty cents for one, a half-dollar for the other.
: 같은 것을 달리 말하다. (50센트는 $\frac{1}{2}$ 달러이다.)

▷ behind the eight ball : 불리한 입장에서

▷ eight ball : 당구에서 8 이라고 쓴 검은 공

▷ once in a blue moon
: 매우 드물게, 좀처럼~않는, seldom

참고 American Heritage 사전에 의하면, 블루문이란,

Chapter 6 영어에서 나타나는 수학용어

1. 같은 달(월)에 뜨는 두 번째 보름달 또는 4번의 보름달이 있는 석 달 캘린더 시즌의 3번째 보름달
2. 일상적으로는 꽤 오랜 기간으로 풀이하고 있으며, 그 어원으로는, 화산 폭발 등으로 인해 대기 중에 있는 상당량의 먼지가 달을 가려 이에 의해 달이 푸르게 나타나는데, 이 달의 출현이 아주 뜸한 데서 '오랜 기간'으로 유래한 것으로 추정함.

▷ one-man band
: 혼자서 여러 악기를 동시에 연주하는 거리의 악사

▷ second to none : 아무것에도 뒤지지 않는, 1등인

▷ He [she] is number one.

☞ He is number one in that field. : 그는 그 분야의 일인자다.

▷ seventh heaven : 최고의 행복, 하늘나라

▷ a square peg in a round hole
: 둥근 구멍 속에 네모난 못

▷ a round peg in a square hole
: 네모난 구멍 속에 둥근 못
: 못의 둘레를 꽉 잡아주지 못해서 쉬 빠진다. 적임자(適任者)가 되지 못한다, 부적격자

▷ one-armed bandit : 슬롯머신. 도박 기계. '외팔의 도적'(레버(lever)를 팔로 보고 붙인 명칭)

▷ at sixes and sevens : 난잡하게, 혼란하여

▷ hit for six : 대성공을 거두다.

▷ six feet under = dead and buried : 죽다.

(미국에서는 사람이 죽으면 관을 땅속 6 피트 아래 묻기 때문에)

▷ a dime a dozen : 흔해 빠진 것, 쉽게 얻을 수 있는 것

▷ a penny saved is a penny earned
: 한 푼을 아끼면 한 푼을 버는 것이다

▷ a piece of cake
: 아주 쉽게 이룰 수 있는 것, 누워서 떡먹기

▷ baker's dozen
: 빵집의 한 다스, 숫자 13을 의미. 원래 한 다스는 12개이지만 빵집에서 한 다스를 사면 빵을 하나 더 주어 13개가 된다.

▷ don't count your chickens before they hatch
: 김칫국부터 먼저 마시지 마라.

▷ don't put all your eggs in one basket
: 계란을 한 바구니에 전부 담지 말라, 한 가지 일에 모든 걸 걸지 말라.

▷ eighty six
: 어떠한 것을 더 이상 이용할 수 없거나 쫓겨난다는 의미

> **참고** 뉴욕의 Bedford 86번가에는 밀주를 팔았던 곳이 있는데 손님이 취하면 쫓아 냈다고도 하며, 술을 팔 때 물을 14% 탄 즉, 86%의 술을 팔았다고도 한다. 뉴욕의 레스토랑에 테이블이 85개뿐이라 86번 째 손님은 입장을 못 한다고 함.

▷ go the extra mile
: 정해진 목표 이상의 임무를 수행하는 의미

Chapter 6 영어에서 나타나는 수학용어

▷ high five
: 축하 혹은 기쁨의 표시로 서로의 손바닥을 마주치는 행위
▷ give me five! : 서로의 손뼉을 치자는 의미
▷ if it's not one thing, it's another : 하나가 잘못되면 또 다른 하나가 잘못되고, 또 다른 하나가 잘못되고.
▷ last but not least : 마지막에 말하는 것이지만 가장 중요한 것을 말할 때 앞의 말들보다 더 중요한 것을 마지막에 쓴다.
▷ Rome was not built in one day : 로마는 하루아침에 건설되지 않았다.
▷ rule of thumb : 어림잡아 계산한다

참고 중세 시대엔 엄지손가락보다 굵지 않은 회초리로 자신의 아내를 매질할 수 있는 것이 합법적이라는 것에서 유래되었다고 한다.

▷ the whole nine yards : 모든 것. 전부
▷ A stitch in time saves nine.
: 제 때의 한 바늘은 아홉 바늘의 수고를 덜어준다.
▷ third times a charm
: 두 번의 실패이후 세 번 째에 성공하는 것
▷ twenty three skidoo : 떠나다, 달아나다
▷ zero tolerance : 제로 관용 정책
(범죄 혹은 법을 어기는 행위가 크든 작든 절대 너그럽게 봐주지 않는다는 뜻)
▷ "영(零)"과 관련된 영어 단어

☞ zero : 영(零)
☞ cipher / cypher (영국식) : 숫자의 영(0)
☞ naught[nɔ:t] / nought : 영(零)
☞ aught[ɔ:t] : 영(零), 제로(nought), 무(無), nothing
☞ null : 영(0), 아무것도 없는, 공집합의
☞ nothing : 영(零)

(예) He is a mere cipher. : 그는 보잘 것 없는 사람이다.
☞ a number of three ciphers : 세 자리의 수
☞ get a naught : 영점을 받다
☞ null and void : 법률상 무효
☞ win by three to nothing : 3 대 0 으로 이기다

▷ half blood : 아버지 또는 어머니가 다른 형제자매관계
▷ half brother, half sister
▷ better half, other half : 남편 또는 아내, 배우자
▷ penultimate step : next to last step
☞ penultimate : 어미에서 둘째의 (음절의); 끝에서 둘째의 것.
▷ have one too many = drink too much alcohol
▷ polymath = polyhistor : 박학자, 박식가
▷ Too many cooks spoil the broth.
: 요리사가 많으면 요리를 망친다. 사공이 많으면 배가 산으로 간다.
▷ a million and one = very many : 매우 많은

Chapter 7 제2외국어 수학 학습 자료들

수학 이외의 언어로 작성된 교재 또는 논문의 경우에는 만국 공통어라는 수식과 기호 등을 보면 고등학교에서 제2외국어로 배운 기본적인 지식을 이용하여 어느 정도 내용을 감을 잡을 수 있다. 이 장에서는 몇몇 외국어에 대한 논문을 검색하기 위한 요령을 익히는 수준이며 문법적인 것을 피하였다.

7.1 중국어 관련 수학

중국어 논문은 수식과 비록 번체이기는 하지만 같은 의미를 가지는 한자어가 등장하기 때문에 감으로 내용을 상당분량 잡을 수 있다. 그러나 정확한 이해를 위해서는 추가적인 정보가 필요하다.

7.1.1 중국수학자 : 중국어 수학책을 학습할 때 자주 나타나는 수학자 이름들

수학자	(한자 : 한자의 한국식 음역)
Abel	(阿贝尔 : 아패이)
Bernoulli	(泊努利 : 박노리)
Cauchy	(柯西 : 가서)
Chebyshev	(切比雪夫 : 절비설부)
Cramer	(克莱姆 : 극래모)
D'Alembert	(达朗贝尔 : 달랑패이)
De Moivre	(棣莫弗 : 체막불)
Euler	(歐拉 : 구랍)
Fourier	(傅立叶 : 부립협)
Gauss	(高斯 : 고사)
Green	(格林 : 격림)
Kronecker	(克朗涅克爾 : 극랑열극이)
L'Hospital	(洛必达 : 락필달)
Lagrange	(拉格朗日 : 랍격랑일)
Laplace	(拉普拉斯 : 랍보랍사)
Leibniz	(来佑尼茨 : 래포니자)
Lindberg	(林德伯格 : 임덕백격)
Maclaurin	(马克勞林 : 마극노림)
Newton	(牛顿 : 우돈)
Poisson	(泊松 : 박송)
Rolle	(罗尔 : 라이)
Simpson	(辛卜生 : 신복생)
Stokes	(斯托克斯 : 사탁극사)
Taylor	(泰勒 : 태륵)
Weierstrass	(外爾斯特拉斯 : 외이사특랍사)

Chapter 7 제2외국어 수학 학습 자료들

7.1.2 중국어 수학용어

拓撲(탁박)	토폴로지(topology)
方程(방정)	방정식(方程式)
维恩图解(유은도해)	벤다이어그램(Venn diagram)
微分方程(미분방정)	미분방정식(differential equation)
派(파)	pi, 원주율
除以(제이)	to divide by
二等边的(이등변적)	isosceles
等边(등변)	equilateral
中数(중수)	median
众数(중수)	mode
正比(정비)	proportion
反比(반비)	inverse ratio
乘(승)	to multiply
4 加 5 得 9.	4 plus 5 is 9. 또는 4 and 5 is 9.
4 加 5 等 于 9.	4 added to 5 equals 9.
5 减 4 得 1.	5 minus 4 is 1.
8 减 3 等 于 5.	8 minus 3 equals 5.
3 减 4 得 负 1.	3 minus 4 is minus 1.
10 减 15 得 负 5.	15 subtracted from 10 is a minus 5.
9 减 4 得 5.	4 subtracted from 9 is 5.
9 减 4 等 于 5.	4 subtracted from 9 equals 5.
9 乘 以 5 得 45.	9 times 5 is 45. 또는 9 multiplied by 5 is 45.
10 除 以 2 得 5.	10 divided by 2 is 5.
89 除 以 3 得 29 点 667.	89 divided by 3 is 29.667.
18 除 以 6 得 3.	18 over 6 is 3.
一 亿 是 一 百 万 的 一 百 倍.	A hundred million is one hundred times a million.

Chinese Numbers

數符號	中文小寫	中文大寫
0	〇	零
1	一	壹
2	二	貳
3	三	叁(參)
4	四	肆
5	五	伍
6	六	陆(陸)
7	七	柒
8	八	捌
9	九	玖
10	十	拾
100	百	佰
1000	千	仟

7.1.3 중국어 수학책이나 논문을 이용할 때

1. 중국어 명사에 的을 붙이면 형용사가 되고 地를 붙이면 부사가 된다. 的과 地 둘은 중국어에서는 발음이 같다.

严格(엄격)	엄격 rigor
严格的(엄격적)	엄격한 rigorous
严格地(엄격지)	엄격하게 rigorously
困难(곤난)	곤란 difficulty
困难的(곤난적)	곤란한 difficult
困难地(곤난지)	곤란하게 with difficulty
直接(직접)	직접 directness
直接的(직접적)	직접적인 direct
直接地(직접지)	직접적으로 directly

Chapter 7 제2외국어 수학 학습 자료들

2. 다음은 부정어를 만드는 역할을 한다.
   不(부), 无(무), 未(미), 非(비)

3. 행동을 나타내는 한자어

了(료)	완료된 행동을 나타냄
着(착)	진행되는 행동을 나타냄
过(과)	경험된 행동을 나타냄

다음은 중국어로 작성된 책이나 논문을 이용하는 데 도움이 되도록 중국어 간체를 이용하여 엮어 놓았다.

한자(음역)	의미
大(대)	big
小(소)	small
是(시)	be
有(유)	have, exist
和(화)	and
或(혹)	or
向(향)	towards
数(수)	number
点(점)	point
线(선)	line
量(량)	quantity
值(치)	value
边(변)	side
数学(수학)	mathematics
同时(동시)	simultaneous
向量(향량)	vector

한자(음역)	의미
定向(정향)	orientation
三角形(삼각형)	triangle
定义(정의)	definition
引理(인리)	lemma
定理(정리)	theorem
推论(추론)	corollary
使得(사득)	so that
只要(지요)	so long as
人人(인인)	everybody
事事(사사)	everything
处处(처처)	everywhere
仅仅(근근)	only
常常(상상)	often
大小(대소)	magnitude
多少(다소)	how much ?
反正(반정)	in any case
左右(좌우)	or thereabouts
矛盾(모순)	contradiction
限(한)	limit
极(극)	extreme
极限(극한)	limit
表示(표시)	representation
代换(대환)	substitution
变换(변환)	transformation
应用(응용)	application
对应(대응)	corresponding
条件(조건)	condition
代数(대수)	algebra
函数(함수)	function
微分(미분)	differential
积分(적분)	integral
数学家(수학가)	mathematician

Chapter 7 제2외국어 수학 학습 자료들

한자(음역)	의미
科学家(과학가)	scientist
子集(자집)	subset
子区间(자구간)	subinterval
子空间(자공간)	subspace
半圆(반원)	semi-circle
半稳定(반온정)	semi-stable
半轴(반축)	semi-axis
任何(임하)	any
任意(임의)	arbitrary
某一(모일)	some, certain
一切(일절)	all
各(각)	each
每一(매일)	every
对任何 c (대임하 c)	for any c
对于任意 x (대우임의 x)	for arbitrary x
某一 t 值(모일 t 치)	some values (of) t
一切系数(일절계수)	all coefficients
在各邻域(재각린역)	in each neighborhood
从每一直线(종매일 직선)	from every [straight] line
其他(기타)	other
另外(령외)	other
别(별)	other
不同(부동)	different, not same
异(이)	different
反(반)	opposite
其他形式(기타형식)	other forms
另外方法(령외방법)	other methods
别的证明(별적증명)	another proof
异号(이호)	opposite sign
反定向(반정향)	opposite orientation
不同的想图(부동적상도)	a different phase portrait
两条线(량조선)	two lines

한자(음역)	의미
四张平面(사장평면)	four planes
一件事(일건사)	a matter
三个奇点(삼개기점)	three singular points
至少一个极限环(지소일개극한배)	at least one limit cycle
二十三个问题(이십삼개문제)	twenty-three problems
最多两个焦点(최다량개초점)	at most two foci
第一类(제일류)	the first kind
第二情况(제이정황)	the second case
第四象限(제사상한)	the fourth quadrant
这(저)	this
那(나)	that
哪(나)	which? what?
这件事(저건사)	the matter
那个问题(나개문제)	that problem
哪条线?(나조선?)	which line?
这些(저사)	these
那些(나사)	those
哪些?(나사?)	which? what?
这些事(저사사)	these matters
那些问题(나사문제)	those problems
一些注意(일사주의)	some remarks
这里(저리)	here
那里(나리)	there
哪里(나리)	where
这儿(저인)	here, now
那儿(나인)	there, then
哪儿(나인)	where
假设(가설)	assume
几(궤)	a few, how many?
整(정)	whole
几个方法(궤개방법)	several methods
在整张平面(재정장평면)	in (the) whole plane

Chapter 7 제2외국어 수학 학습 자료들

한자(음역)	의미
我(아)	I
我们(아문)	we
你(니)	you(단수)
你们(니문)	you(복수)
他(타)	he
她(저)	she
关于他的猜测(관우타적시측)	concerning his conjecture
在她的文中(재저적문중)	in her article
因为它的困难(인위타적곤난)	because of its difficulty
其中(기중)	where, in which
其中数 k 负(기중수 k 부)	where (the) number k (is) negative
开集(개집)	open set
闭曲线(폐곡선)	closed curve
奇点(기점)	singular point
复数(복수)	complex number
实部(실부)	real part
正号(정호)	positive sign
下界(하계)	lower bound
上限(상한)	upper limit
右边(우변)	right (hand) side
单根(단근)	simple root
初值(초치)	initial value
左侧(좌측)	left (hand) side
高阶(고계)	higher order
二次系统(이차계통)	quadratic system
齐次坐标(재차좌표)	homogeneous coordinates
幂级数(멱급수)	power series
周期解(주기해)	periodic solution
独立变量(독립변량)	independent variable
足够小(족구소)	sufficiently small
渐近稳定(점근온정)	asymptotically small
充要条件(충요조건)	necessary and sufficient condition

원서읽기를 위한 수학용어사전

한자(음역)	의미
变数的更换(변수적경환)	change of variables
点的邻机(점적린궤)	neighborhood of a point
坐标的原点(좌표적원점)	origin of coordinates
方程组(방정조)	system of equations
相交角(상교각)	angle of intersection
目的(목적)	aim, purpose
地方(지방)	place
因为(인위)	because
由(유)	because of
选取(선취)	choose
小(소)	small
较小(교소)	smaller
最小(최소)	smallest
多(다)	many, much
较多(교다)	more
最多(최다)	most
好(호)	good
更好(경호)	better
最好(최호)	best
仅(근)	only
只(지)	only
相当(상당)	quite
颇(파)	rather
很(흔)	very, quite
非常(비상)	very
甚(심)	very, extremely
极(겁)	extremely
仅一个(근일개)	only one
只有两个(지유량개)	only two
想当困难(상당곤난)	quite difficult
颇复杂(파복잡)	rather complicated
很有用(흔유용)	very useful

Chapter 7 제2외국어 수학 학습 자료들

한자(음역)	의미
非常容易地(비상용역지)	very easily
甚高频(심고빈)	very high frequency
极小值(극소치)	minimum [value]
正根(정근)	positive root
不等(부등)	unequal
不全(불전)	incomplete
不定积分(부정적분)	indefinite integral
不但~ 而且~ (부단 ~ 이차 ~)	not only ~ but also~
无界的(무계적)	unbounded
无限的(무한적)	infinite
未知量(미지량)	unknown
非线性(비선성)	nonlinear
他有(타유)	he has
他们有(타문유)	they have
表示(표시)	express, representation
代换(대환)	substitute, substitution
变换(변환)	vary, transformation
应用(응용)	apply, application
注意(주의)	take note of, remark
正规化(정규화)	normalize
不可能的(불가능적)	impossible
简化(간화)	simplify
是(시)	do(강조할 때), be
有(유)	have, possess
要(요)	want, need, important
来(래)	come
去(거)	go
到(도)	arrive
走(주)	leave
进(진)	enter
出(출)	exit
在(재)	exist

원서읽기를 위한 수학용어사전

한자(음역)	의미
给(급)	give
对(대)	answer
跟(근)	follow
能(능)	can
会(회)	be able to
应当(응당)	should
得(득)	must
能够(능구)	be capable of
可以(가이)	may
应该(응해)	ought to
必须(필수)	have to
可(가)	can, –able
设(설)	suppose
假(가)	assume
如果(여과)	if
若(약)	if
当(당)	when, if
虽然(수연)	although
除非(제비)	unless
当且仅当(당차근당)	if and only if
谁(수)	who?(의문문일 때), whoever(의문문이 아닐 때)
什么(십요)	what?(의문문일 때), whatever(의문문이 아닐 때)
哪儿(나인)	where?(의문문일 때), wherever(의문문이 아닐 때)
几(궤)	how many?(의문문일 때), a few(의문문이 아닐 때)
怎么(즘요)	how?
为什么(위십요)	why?
什么时候(십요시후)	when?
给(급)	give
叫(규)	call
让(양)	let

Chapter 7 제2외국어 수학 학습 자료들

한자(음역)	의미
首先(수선)	first
现在(현재)	now
今(금)	now
其次(기차)	next
然后(연후)	then
后来(후래)	later
最后(최후)	finally
最近(최근)	recently
而且(이차)	moreover
然而(연이)	however
但(단)	but
如此(여차)	thus
因此(인차)	hence
因为(인위)	because
为了(위료)	in order to
又(우)	again
此外(차외)	in addition
不过(불과)	nevertheless
可是(가시)	but
由此(유차)	thus
于是(우시)	consequently
由于(유우)	since
所以(소이)	therefore
则(칙)	then
当然(당연)	of course
其实(기실)	in fact
即(즉)	that is
例如(례여)	for example
特别(특별)	in particular
同样(동양)	similarly
下面(하면)	in the following
显见(현견)	obviously

한자(음역)	의미
实际上(실제상)	actually
再(재)	again
也许(야허)	perhaps
根据(근거)	according to
另一方面(령일방면)	on the other hand

7.2 러시아어 관련 수학

7.2.1 러시아어 알파벳 (33개)

 러시아어 논문을 이해하기 위해서는 알파벳 33개의 발음을 알면 수학 용어는 쉽게 따라 잡을 수 있다. 그러나 추가적으로 간단한 문법과 형용사, 부사와 명사의 지식이 필요하다. 대체로 '러시아어-영어 수학용어 사전'을 보면 상당히 같은 발음을 가지는 수학용어가 많은 것에 놀라게 된다.

 러시아어를 나타내는 문자는 키릴 문자라고 하며, 다른 문자보다 더욱 그리스 문자에 가까운 형태를 하고 있다. 예를 들어 그리스어에서 Γ, Δ, Λ, Π, Ρ 등이 그렇다. 키릴 문자는 대문자와 소문자가 대부분 같고 크기만 다르기 때문에 기억하기가 쉽다. 다음 표에서 러시아 대문자, 소문자 33개를 순서대로 나타내었다.

 예전에는 러시아에서 출판된 수학관련 논문들 중에 다수의 논문들은 AMS 등을 통해서 몇 개월 안에 영어로 번역되었으며, 간혹 번역이 되지 않는 논문 중에서 제목과 abstract 를 검색하여 필요한 논문을 구할 수 있으면 될 것 같다.

Chapter 7 제2외국어 수학 학습 자료들

대문자	소문자	발음	대문자	소문자	발음
А	а	[a/아]	Р	р	[r/에르]
Б	б	[b/베]	С	с	[s/에쓰]
В	в	[v/붸]	Т	т	[t/떼]
Г	г	[g/게]	У	у	[u/우]
Д	д	[d/데]	Ф	ф	[f/에프]
Е	е	[je/예]	Х	х	[x/하]
Ё	ё	[jo/요]	Ц	ц	[ts/쩨]
Ж	ж	[zh/줴]	Ч	ч	[ch/쳬]
З	з	[z/제]	Ш	ш	[sh/샤]
И	и	[i/이]	Щ	щ	[shy/시챠]
Й	й	[j/이 끄라뜨꼬이]	Ъ	ъ	[뜨뵤르드이 즈나크]
К	к	[k/까]	Ы	ы	[i/의]
Л	л	[l/엘]	Ь	ь	[먀흐끼이 즈나크]
М	м	[m/엠]	Э	э	[e/에]
Н	н	[n/엔]	Ю	ю	[ju/유]
О	о	[o/오]	Я	я	[ja/야]
П	п	[p/뻬]			

7.2.2 러시아 수학자 : 러시아어 수학책을 학습할 때 자주 나타나는 수학자 이름들

Abel	Абель
Adams	Адамс
Agnesi	Аньези
Airy	Эйри
Alexander	Александер
Archimedes	Архимед
Argand	Арган
Artin	Артин
Atiyah	Атья
Banach	Банах
Bernside	Бертран
Bernoulli	Бернулли
Bessel	Бессель
Betti	Бетти
Binet	Бине

Birkhoff	Биркгоф
Bonnet	Бонне
Bool	Буль
Borel	Борель
Bourbaki	Бурбаки
Brouwer	Брауэр
Brown	Броун, Браун
Cantor	Кантор
Cardano	Кардано
Cauchy	Коши
Cayley	Кэли
Ceva	Чева
Chebyshev, Tschebycheff	Чебышёв
Clairaut	Клеро
Cohen	Коэн
d'Alembert	Даламбер
Darboux	Дарбу
Courant	Курант
Cramer	Крамер
De Morgan	Морган
Dedekind	Дедекинд
Desargues	Дезарг
Descartes	Декарт
Dirac	Дирак
Dirichlet	Дирихле
Einstein	Эйнштейн
Eisenstein	Эйзенштейн
Eratosthenes	Эратосфен
Erdös	Эрдёш

Chapter 7 제2외국어 수학 학습 자료들

Euclid	Эвклид
Euler	Эйлер
Fermat	Ферма
Fermi	Ферми
Fibonacci	Фибоначчи
Field	Филдс
Fourier	Фурье
Galois	Галуа
Gateaux	Гато
Gauss	Гаусс
Frechet	Фреше
Frenet	Френе
Frobenius	Фробениус
Fubini	Фубини
Gödel	Гёдель
Goldbach	Гольдбах
Goursat	Гурса
Harr	Хаар
Hadamard	Адамар
Hahn	Хан
Hamilton	Хэмнлтон
Hardy	Харди
Grasssman	Грассман
Green	Грин
Hasse	Хассе
Hausdorff	Хаусдорф
Heine	Гейне
Hermit	Эрмит
Hesse	Гессе

원서읽기를 위한 수학용어사전

Hilbert	Гильберт
Hölder	Гёльдер
Hooke	Гук
Horner	Горнер
Jacobi	Якоби
Jordan	Жордан
Kuratowski	Куратовский
Lagrange	Лагранж
Laguerre	Лагерр
Laplace	Лаплас
Kronecker	Кронекер
Legendre	Лежапдр
Leibniz	Лейбниц
L'Hospital	Лопиталь
Lie	Ли
Lindelöf	Линделёфа
Liouville	Лиувилль
MacLane	Маклейн
MacLaurin	Маклорен
Markov	Марков
Mazur	Мазур
Mittag-Leffler	Миттаг-Леффлер
Minkowski	Минковский
Möbius	Мёбиус
De Moivre	Муавр
Napier	Непер
Newton	Ньютон
Pascal	Паскаль
Peano	Пеано

Chapter 7 제2외국어 수학 학습 자료들

Penrose	Пенроуз
Pfaff	Пфафф
Poincaré	Пуанкаре
Poisson	Пуассон
Polya	Пойа
Ramanujan	Рамануджан
Rayleigh	Рэлей
Raccati	Риккати
Riemann	Риман
Rolle	Ролль
Russel	Рассел
Schmidt	Шмидт
Schwarz	Шварц
Stieltjes	Стилтьес
Stokes	Стокс
Sylow	Силов
Taylor	Тэйлор
Ulam	Улам
Urysohn	Урысон
Venn	Венн
Waerden, van der	Ван-дер-Варден
Wronski	Вроньский
Yoshida	Иосида
Zermelo	Цермело
Zorn	Цорн

7.2.3 러시아어 명사

러시아어 수(數)에 따른 명사의 복수형태가 달라지므로 수학적인 용어를 검색할 때 명사의 어근을 잘 기억해야 한다. 러시아에는 명사에 남성, 여성, 중성이 있다. 명사의 경우는 하나를 나타내는 단수형과 2, 3, 4를 나타내는 복수형과 5 이상을 나타내는 복수형이 있음을 주의해야 한다. 참고로 폴란드어에서도 명사에서 이와 같은 현상이 나타난다.

▷ parasol (우산, parasol) - 남성형 명사
　하나일 때　　　parasol
　둘,셋,넷일 때　parasole
　다섯 이상일 때 parasoli

▷ karta (카드, card) - 여성형 명사
　하나일 때　　　karta
　둘,셋,넷일 때　karty
　다섯 이상일 때 kartek

▷ radio (라디오, radio) - 중성형 명사
　하나일 때　　　radio
　둘,셋,넷일 때　radia
　다섯 이상일 때 radii

Chapter 7 제2외국어 수학 학습 자료들

7.2.4 러시아어 수학기호와 읽기

$a = b$	a равно b
$a \neq b$	a не равно b
$a \approx b$ 또는 $a \doteqdot b$	a примерно (приближенно) равно b
$a > b$	a больше b ; a больше, чем b
$a < b$	a меньше b ; a меньше, чем b
$x = \infty$	равен бесконечности
$|x| = 1$	одиножды один (будет) один
$a + b = c$	a плюс b равно (будет) c
$35 + 16 = 51$	тридцать пять плюс (прибавить) шестнадцать будет (составит) пятьдесят один (равняется пятидесяти одному)
$a - b = c$	a минус b равно (равняется) c
$64 - 17 = 47$	шестьдесят четыре минус (отнять, вычесть) семнадцать будет (составит) сорок семь (равно сорока семи)
$=$	равняться (чему), равно (чему)
$\equiv$	тожественно равняться (чему), тожественно равно (чему)
$\neq$	не равняться (чему), не равно (чему)
$\approx$ 또는 $\doteqdot$	приближенно равняться (чему), приближенно равно (чему)
$<$	меньше (чего)
$>$	больше (чего)

≤	меньше или равно
≥	больше или равно
≪	значительно меньше 또는 мало по сравнеиию с (чем)
$\|a\|$	абсолютная величина числа a
+	плюс-сложение
−	мннус-вычитание
· 또는 ×	умножиться на (что)-умножение
: 또는 ÷	делиться на (что)-деление
a^m	a в степени m
$\sqrt{}$	квадратный корень
$\sqrt[n]{}$	корень n-ной степени
( ), [ ], { }	скобкн
!	факториал
⊥	перпендикулярно (к чему)
//	параллельно (чему)
#	равно и параллельно
∼	подобно (чему)
△	треугольник
∠	угод(예 : $\angle ABC$)
⌢	дуга(예 : $\overset{\frown}{AB}$)
°	градус(예 : 32°)

Chapter 7 제2외국어 수학 학습 자료들

′	минута (분) (예 : 11′)
″	секунда (초) (예 : 14″)
sin	синус (사인)
cos	косинус (코사인)
tg	тангенс (탄젠트)
ctg	котангенс (코탄젠트)
sc	секанс (시이컨트)
csc	косеканс (코시이컨트)
const	постоянная величина (константа)
lim	предел (лимит)
→	стремится к (чему)
∞	бесконечность
∝	пропорционально
$\sum$	сумма
$\sum_{i=1}^{n}$	сумма, где i изменяется от 1 до n
$f(\ )$	функция от (чего)
Δ	приращение (예 : Δx)
d	диференциал (예 : dx)
$\int$	интеграл
$\int_{a}^{b}$	определенный интеграл от нижнего предела a до верхнего предела b
$\iint$	двойной интеграл

원서읽기를 위한 수학용어사전

$\iiint$	тройной интеграл
$\log b$	логарифм при основании b
lg	десятичной логарифм
ln	натуральный логарифм
$a = c \times d$	a равно c помноженное (умноженное) на d
$1 : 2$	один к двум (отношение один к двум)
$15 : 3 = 5$	пятнадцать, поделенное на три, равно пяти
$12 : 3 = 16 : 4$	отношение двенадцать к трем равно отношению шестнадцать к четрем
$a = \dfrac{b}{c}$	a равно b деленному на c
$a = \dfrac{\frac{b}{c}}{\frac{d}{e}} = \dfrac{be}{cd}$	a равно отношению b поделенного на c к d, поделенному на c, равно (равняется) отношению произведения be к произведению cd
$\dfrac{1 \times 100}{15} = 6\dfrac{2}{3} \approx 6.66$	единица, умноженная на сто и поделенная на пятнадцать составляет шесть и две трети, или риблизительно шесть целых и шестьдесят шесть сотых
x^2	икс квадрат; икс в квадрате; икс в степени два; икс во второй степени

Chapter 7 제2외국어 수학 학습 자료들

x^3	икс куб; икс в кубе; икс в степени три; икс в третьей степени
10^{-6}	десять в степени минус шесть
$\sqrt{a}$	корень квадратный из a
$\sqrt[3]{y}$	корень кувический из a
$\sqrt[4]{c}$	корень четвертой степени из a

7.3 프랑스어 관련 수학

　프랑스는 영어와 같은 26개의 알파벳을 이용하기 때문에 사전을 이용하기에 큰 불편은 없다. 다만 영어와 같은 철자의 단어가 발음에 다른 경우가 많다. 프랑스어 논문을 이해하기 위해서는 고등학교에서 제2외국어로 배우는 정도의 프랑스어 지식과 수학적인 용어에 대한 추가적인 학습이 필요하다. 특히 프랑스어의 시제의 다양성과 그에 따른 특히 불규칙적인 동사의 변화가 커다란 장애물로 나타난다.

7.3.1 정관사와 부정관사

프랑스어 정관사

	남성	여성
단수	le	la
복수	les	les

프랑스어 부정관사

	남성	여성
단수	un	une
복수	des	des

à + le → au

à + les → aux

de + le → du

de + les → des

7.3.2 수식과 수식 표현

(1) 분수

$\dfrac{1}{2}$	demi (별도의 용어이다) 관사까지 포함하여 un demi 또는 une (la) moiti
$\dfrac{1}{3}$	tiers (고어를 사용한다) 관사까지 포함하여 un tiers
$\dfrac{1}{4}$	un quart (고어를 사용한다) un quart
$\dfrac{1}{5}$	un cinqui me
$\dfrac{1}{6}$	un sixi me
$\dfrac{1}{7}$	un septi me
$\dfrac{1}{8}$	un huiti me

Chapter 7 제2외국어 수학 학습 자료들

$\dfrac{1}{9}$	un neuvième
$\dfrac{1}{10}$	un dixième
$\dfrac{2}{3}$	deux tiers
$\dfrac{3}{4}$	trois quarts
$\dfrac{3}{5}$	trois cinquièmes
$\dfrac{7}{10}$	sept dixièmes
$\dfrac{5}{12}$	cinq douzièmes
$\dfrac{1}{100}$	un centième
$\dfrac{9}{100}$	neuf centièmes
$\dfrac{1}{300}$	un trois centième
$1\dfrac{1}{2}$	un et demi
$4\dfrac{3}{4}$	quatre trois quarts
$\dfrac{14}{327}$	quatorze sur trois cent vingt-sept
$\dfrac{356}{545}$	trois cent cinquante-six sur cinq cent quarante-cinq

주의 분모가 5이상이면 분모는 서수로 분자는 기수로 나타낸다. 그리고 분자가 2 이상이면 분모를 복수로 한

다(s를 붙임). 분자는 기수로 표시하고, 분자의 수가 2 이상일 때는 분모의 서수는 복수형을 취한다. 분자 · 분모가 큰 수일 경우는 기수를 사용하여 $\frac{a}{b}$ 를 a sur b 로 읽는다.

(2) 소수

프랑스에서는 소수(fraction décimale)를 나타낼 때 쉼표를 사용한다.

38,4	trente-huit virgule quatre (trente-huit entiers quatre dixi mes)
0,532	z ro virgule cinq trois deux
0,2	z ro virgule deux
8,5	huit virgule cinq
15,41	quinze virgule quarante et un
10%	dix pour cent
16%	seize pour cent

(3) 수학 기호

+	plus
−	moins
±	plus ou moins
×	multipli par

Chapter 7 제2외국어 수학 학습 자료들

$\div$	divisé par
$\dfrac{a}{b}$	a sur b
$=$	égal
$\equiv$	identique
$\neq$	différent de
$<$	inférieur
$\leq$	inférieur ou égal
$\ll$	très inférieur
$>$	supérieur
$\geq$	supérieur ou égal
$\gg$	très supérieur
Σ	sigma de
$\in$	appartient
$\notin$	n'appartient
$\subset$	inclus dans
$\not\subset$	non inclus dans
$\cup$	symbole d'union
$\cap$	symbole d'intersection
$\lvert x \rvert$	valeur absolue de x

(4) 거듭제곱(puissance)과 제곱근

제곱과 세제곱은 서수로 표시한다

원서읽기를 위한 수학용어사전

제곱	la deuxième puissance (= le carré)
세제곱	la troisième puissance (= le cube)
네제곱	la quatrième puissance
3^2	la deuxième puissance (= le carré) de trois
5^4	la quatrième puissance de cinq; cinq à la quatrième puissance 또는 cinq (à la) puissance quatre
4^2	le carré de quatre; quatre au carré; quatre (à la) puissance deux
4^3	le cube de quatre; quatre au cube; quatre (à la) puissance trois
a^n	a à la $n^{ième}$ puissance
$\sqrt{10}$	la racine carrée de dix
$\sqrt[3]{10}$	la racine cubique de dix
$\sqrt[4]{10}$	la racine quatrième de dix
$\sqrt[n]{x}$	la racine $n^{ième}$ de dix

(5) 배수

2배	le double
3배	le triple
4배	le quadruple
5배	le quintuple
6배	le sextuple
7배	le septuple
8배	le octuple
9배	le nonuple
10배	le décuple
100배	le centuple

참고 '몇 배'를 나타낼 때는 fois 를 사용할 수 있다. 예를 들어 2배는 deux fois, 3배는 trois fois 이다.

Chapter 7 제2외국어 수학 학습 자료들

(6) 어림수

대략이라는 어림수를 나타낼 때는 기수의 어미에다 aine 를 붙인다.

대략 8	une huitaine
대략 10	une dizaine
대략 12	une douzaine
대략 15	une quinzaine
대략 30	une trentaine
대략 100	une centaine
대략 1000	une millier

(7) 사칙연산

더하기(addition)

| 2 + 3 = 5 | Deux et [plus] trois font [galent; gale] cinq. |
| 3 + 9 = 12 | Trois et neuf font douze. [font → faire] |

빼기(soustraction)

| 8 − 1 = 7 | Huit moins un font sept. |
| 16 − 5 = 11 | Seize moins cinq gale onze. [gale → galer] |

곱하기(multiplication)

| 6 × 9 = 54 | Six multipli   par neuf   gale cinquante-quatre; 또는 Six fois neuf [Neuf fois six] font cinquante-quatre. |
| 13 × 7 = 91 | Treize fois sept font quatre-vingt-onze. |

나누기(division)

12 ÷ 3 = 4	Douze divisé par trois égale quatre.
60 ÷ 15 = 4	Soixante divisé par quinze égale quatre. [divisé → diviser]

주의

▷ 1은 남성형 un 과 여성형 une 가 있다. 21, 31, 41, 51, … 에서도 1만은 남성과 여성의 구별이 있다. un 은 명사 뒤에 놓이면 여성형으로 쓰이지 않고 항상 un 이 된다.

▷ 21, 31, 41, 51, 61, 71 여섯 개는 접속사 et 로 연결한다. 나머지 100 이하의 복합형은 모두 - (trait d'union)으로 연결한다.

▷ vingt (20) 과 cent (100)의 배수는 복수를 나타내는 s를 붙이지만 뒤에 계속해서 다른 숫자가 따르면 s를 생략한다. 서수에서도 같다.

▷ million(백만), milliard(10억)은 명사이므로 복수형이 된다. 그리고 뒤에 명사가 계속해서 나오면 전치사 de로 연결한다. 예를 들어 300만 프랑은 trois millions de francs 가 된다.

▷ mille(1천)은 복수형이 없다. 예를 들어 4,000은 quatre mille 이다. 그리고 연도를 나타낼 때는 mil을 사용한다.

▷ 프랑스의 프랑스어에서는 숫자 20 을 vingt 이라고 하고 80 을 quatre-vingts (4개의 20)이라는 표현하고, 90 은

Chapter 7 제2외국어 수학 학습 자료들

quatre-vingt-dix (4개의 20 더하기 10)이 된다. 마찬가지로 70은 soixante-dix (60 더하기 10)이 된다.
▷ 프랑스어 수(數)는 1부터 19까지, 20부터 69까지, 70부터 99까지, 100 이상으로 나누어 기억하면 좋다. 먼저 1부터 19까지는 기본으로 암기하면 된다.
▷ 20부터 69까지는 구성방식이 유사하다.
▷ 70부터 99까지는 70+은 60에 10부터 19까지를 더하고, 90+ 는 80에 10부터 19 까지 더하는 형태가 된다. 100 이상은 1부터 99까지 방식을 이용하여 구한다.
▷ 천단위 구별은 point (.)를 사용하고 소수점은 virgule (,)을 사용한다. virgule과 virgula는 각각 프랑스어와 포르투갈어에서 comma 를 말한다.

프랑스어 기수사

0 zéro	10 dix	20 vingt	80 quatre-vingts
1 un(남) une(여)	11 onze	21 vingt et un	90 quatre-vingt-dix
2 deux	12 douze	22 vingt-deux	99 quatre-vingt-dix-neuf
3 trois	13 treize	30 trente	100 cent
4 quatre	14 quatorze	31 trente et un	101 cent un
5 cinq	15 quinze	32 trente-deux	300 trois cents
6 six	16 seize	40 quarante	1000 mille
7 sept	17 dix-sept	50 cinquante	10000 dix mille
8 huit	18 dix-huit	60 soixante	12400 douze mille quatre cents
9 neuf	19 dix-neuf	70 soixante-dix	백만 : million

7.4 스페인어 관련 수학

　스페인어는 몇 개의 발음을 제외하고는 우리나라 사람들이 우리말을 사용할 때처럼 발음하기도 쉽고 거의 예외 없이 발음하기 때문에 편하다. 문제는 우리가 국내에서 스페인어로 작성된 수학 논문을 자주 접할기회가 없다는 사실이다.

　스페인어는 30개의 알파베또(Alfabeto)가 있지만 영어에 없는 4개인 Ch (소문자는 ch : C와 D 사이), LL (소문자는 ll : L과 M 사이), Ñ (소문자는 ñ : N과 O 사이), rr (대문자 없음 : R과

Chapter 7 제2외국어 수학 학습 자료들

S 사이)만 주의하면 영어 단어 찾듯이 스페인어 사전을 거의 불편 없이 이용할 수 있다.

7.4.1 스페인어 수사 (los numerales)

스페인어 기수사

0 cero	10 diez	20 veinte	80 ochenta
1 uno(남) una(여)	11 once	21 veintiuno 또는 veinte y uno	90 noventa
2 dos	12 doce	22 veintidós	99 noventa y nueve
3 tres	13 trece	30 treinta	100 cien [다른 숫자 앞에서는 ciento-]
4 cuatro	14 catorce	31 treinta y uno	101 ciento uno
5 cinco	15 quince	32 treinta y dos	199 ciento noventa y nueve
6 seis	16 dieciséis 또는 diez y seis	40 cuarenta	200 doscientos(as)
7 siete	17 diecisiete 또는 diez y siete	50 cincuenta	900 novecientos(as)
8 ocho	18 dieciocho 또는 diez y ocho	60 sesenta	999 novecientos noventa y nueve
9 nueve	19 diecinueve 또는 diez y nueve	70 setenta	백만 : un millón 이백만 : dos millones

▷ 기수사에서 1, 21, 31, … 와 이백 이상의 수는 남성, 여성을 구분한다.

    21: veintiuno, veintiuna

▷ uno는 모음 앞에서 un으로 줄여 쓴다.

▷ 서수사에서 모든 숫자는 남성, 여성을 구분한다.
  (예) primero, primera

스페인어에서 "백만"의 경우에는 명사로만 쓰이기 때문에 뒤에 다른 명사가 나올 경우에는 "~ millones de + 명사"의 형태가 된다. 예를 들어

백만(1,000,000)	un millón
이백만(2,000,000)	dos millones
삼백만(3,000,000)	tres millones
천만(10,000,000)	diez millones (백만이 열개)
1억(100,000,000)	cien millones (백만이 백개)
10억(10,000,000,000)	diez mil millones (백만이 천개)

주의 스페인어에는 10억을 나타내는 단어가 따로 없다.

7.4.2 스페인어 관사

외국어 원서를 읽기 위해서는 먼저 관사를 정리하면 명사로 이루어진 수학용어는 영어와 유사한 것이 많으므로 빨리 감을 잡을 수 있다.

정관사		남성	여성
	단수	el	la
	복수	los	las

부정관사		남성	여성
	단수	un	una
	복수	unos	unas

Chapter 7 제2외국어 수학 학습 자료들

7.4.3 스페인어 분수와 배수

$\frac{1}{2}$, 절반	un medio = la mitad (참고) medio : 절반의
$\frac{1}{3}$	un tercio = una tercera parte
$\frac{2}{3}$	dos tercios
$\frac{1}{4}$	un cuarto = una cuarta parte
$\frac{1}{5}$	un quinto = una quinta parte
$\frac{3}{4}$	tres cuartos
$\frac{1}{5}$	un quinto = una quinta parte
$\frac{2}{5}$	dos quintos
$\frac{3}{5}$	tres quintos
$\frac{5}{7}$	cinco s ptimos
$\frac{5}{6}$	cinco sextos
$\frac{1}{10}$	un d cimo
$1\frac{3}{5}$	uno y tres quintos
$2\frac{5}{8}$	dos y cinco octavos

2배의, 2중의	doble 또는 duplo
3배의, 3중의	triple
한번	una vez
다시 한번	otra vez
두 번	dos veces
두 배 크다	dos veces más grande
한 짝	un par
한 다스	una docena

7.4.4 스페인어 사칙연산

(1) 더하기(más)

3 + 12 = 15	Tres más doce son quince.
5 + 3 = 8	Cinco y tres son ocho.

(2) 빼기(menos)

27 − 13 = 14	Veintisiete menos trece son catorce.
5 − 3 = 2	Cinco menos tres son dos.

Chapter 7 제2외국어 수학 학습 자료들

(3) 곱하기(por)

6 × 7 = 42	Seis por siete son cuarenta y dos.
5 × 3 = 15	Cinco por tres son quince.

(4) 나누기(dividido por)

24 ÷ 4 = 6	Veinticuatro dividido por cuatro son seis.
15 ÷ 3 = 5	Quince dividido por tres son cinco.

7.4.5 스페인어 소수

스페인어에서는 소수점에 점(.) 대신에 콤마(,)를 사용한다. 그리고 읽을 때도 점(punto)이라고 읽지 않고 콤마(coma 또는 간단히 con)라고 읽는 점이 특이하다. 그리고 소수점 이하 둘째 짜리까지는 소수점이하 십 단위로 묶어서 읽는다.

0,7	cero coma siete	(영 콤마 칠)
4,5	cuatro coma cinco	(4 콤마 오)
3,14	tres coma catorce	(3 콤마 십사)
12,4	doce coma cuatro	(12 콤마 사)

7.5 독일어 관련 수학

7.5.1 독일어 관사

정관사는 지시 또는 한정을 나타내는 데 쓰이고, 부정관사는 특정되지 않은 사물을 나타내는 데 쓰인다.

정관사

	남성단수	여성단수	중성	복수
주격　　(~은,는)	der	die	das	die
소유격　(~의)	des	der	des	der
간접목적격 (~에게)	dem	der	dem	den
직접목적격 (~을)	den	die	das	die

부정관사 (영어의 a에 해당하는 것으로 복수형은 없다.)

	남성단수	여성단수	중성	복수
주격　　(~은,는)	ein	eine	ein	-
소유격　(~의)	eins	einer	eines	-
간접목적격 (~에게)	einem	einer	einem	-
직접목적격 (~을)	einen	eine	ein	-

7.5.2 독일어 수학기호와 읽기

$a + b = c$　　a plus b ist (gleich) c ; a und b ist c

$a - b = c$　　a minus b ist (gleich) c ; a weniger b ist c

$a \times b = c$
또는 $a \cdot b = c$　　a mal b ist (gleich) c

Chapter 7 제2외국어 수학 학습 자료들

$a : b = c$ 또는 $a/b = c$	a (geteilt) durch b ist (gleich) c
$a \neq b$	a ist nicht gleich b
$a \approx b$	a ist ungefähr (gleich) b
$a \equiv b$	a ist identisch (gleich) b
$a > b$	a ist größer als b
$a \geq b$	a ist größer oder gleich b
$a < b$	a ist kleiner als b
$a \leq b$	a ist kleiner oder gleich b
$a \gg b$	a ist sehr groß gegen b
$a \ll b$	a ist sehr klein gegen b
$n!$	n Fakultät
$f(x)$	Funktion von x
$(a+b)^n$	Klammer auf, a plus b, Klammer zu, hoch n
a^n	a hoch n ; n -te Potenz von a
$\sqrt{a}$	Wurzel aus a ; Quadratwurzel aus a ; zweite Wurzel aus a
$\sqrt[3]{a}$	Kubikwurzel aus a ; dritte Wurzel aus a
$\sqrt[n]{a}$	n -te Wurzel aus a
$\int_a^b f(x)\,dx$	Integral von $f(x)$ von a bis b

7.5.3 숫자읽기 - 분수와 소수

$\frac{1}{2}$	ein halb
$1\frac{1}{2}$	eineinhalb; anderthalb
$2\frac{1}{2}$	zweieinhalb
$\frac{1}{3}$	ein Drittel
$\frac{2}{3}$	zwei Drittel
$\frac{1}{4}$	ein Viertel
$\frac{3}{4}$	drei Viertel
$5\frac{5}{6}$	fünf fünf Sechstel
$\frac{1}{12}$	ein Zwölftel
$\frac{1}{100}$	ein Hundertstel
$\frac{1}{1000}$	ein Tausendstel

소수에서 소수점은 콤마로 나타내고 소수점 이하의 수는 보통 한자리씩 읽는다.

18,21	achtzehn Komma zwei eins
21,18	einundzwanzig Komma ein acht

Chapter 7 제2외국어 수학 학습 자료들

독일어 기수사

0 null	14 vierzehn	60 sechzig
1 ein	15 fünfzehn	70 siebzig
2 zwei	16 sechzehn	80 achtzig
3 drei	17 siebzehn	90 neunzig
4 vier	18 achtzehn	99 neunundneunzig
5 fünf	19 neunzehn	100 hundert
6 sechs	20 zwanzig	101 einhundertein
7 sieben	21 einundzwanzig	113 einhundertdreizehn
8 acht	22 zweiundzwanzig	123 einhundertdreiundzwanzig
9 neun	30 dreissig	300 dreihundert
10 zehn	31 einunddreissig	1000 eintausend
11 elf	32 zweiunddreissig	10000 zehntausend
12 zwölf	40 vierzig	
13 dreizehn	50 fünfzig	

주의 (1) 2부터 9까지 수에 zig 를 붙여서 두 자리수의 기수를 만든다.

(2) 일의 자리 숫자를 먼저 읽거나 쓰고 '그리고(und)'를 넣고 십의 자리 숫자를 읽거나 쓴다. 예를들어 '24' 는 '4 and 20' 으로 'vier + und + zwanzig' 을 붙여서 쓴 vierundzwanzig 가 된다.

7.6 Hebrew

문자 (letter)	이름 (name)	음역 (transliteration)	문자 (letter)	이름 (name)	음역 (transliteration)
א	aleph	—or'	מ, ם	mem	m
ב	beth	b,v	נ, ן	nun	n
ג	gimel	g	ס	samekh	s
ד	daleth	d	ע	ayin	'
ה	he	h	פ, ף	pe	p,f
ו	vav	v,w	צ, ץ	sadi	ṣ
ז	zayin	z	ק	koph	ḵ
ח	cheth	ḥ	ר	resh	r
ט	tech	t	שׁ	shin	sh,š
י	yod	y,j,i	שׂ	sin	ś
כ, ך	kaph	k,kh	ת	tav	t
ל	lamed	l			

주의 단어(word)의 마지막에 올 때는 두 개있는 경우는 뒤에 있는 문자를 사용한다.

7.7 GREEK alphabet

처음에는 27문자였으나 3개가 없어지고 현재 24문자가 남아 있다. epsilon 다음에 stigma, pi 다음에 qoppa, omega 다음에 sampi 가 없어진 문자이다.

Chapter 7 제2외국어 수학 학습 자료들

문자(letter)		이름(name)	음역(transliteration)	대응하는 숫자
A	α	alpha	a	1
B	β	beta	b	2
Γ	γ	gamma	g	3
Δ	δ	delta	d	4
E	ϵ	epsilon	e	5
	ς	stigma	(현재 없어진 문자)	6
Z	ζ	zeta	z	7
H	η	eta	e 또는 ē	8
Θ	θ	theta	th	9
I	ι	iota	i	10
K	κ	kappa	k	20
Λ	λ	lambda	l	30
M	μ	mu	m	40
N	ν	nu	n	50
Ξ	ξ	xi	x	60
O	o	omicron	o	70
Π	π	pi	p	80
	$\qoppa$	qoppa	(현재 없어진 문자)	90
P	ρ	rho	r	100
Σ	σ, s	sigma	s (단어의 끝에 올 때는 소문자는 s를 사용한다)	200
T	τ	tau	t	300
Υ	υ	upsilon	y	400
Φ	ϕ	phi	ph	500
X	χ	chi	ch, kh	600
Ψ	ψ	psi	ps	700
Ω	ω	omega	o 또는 ō	800
		sampi	(현재 없어진 문자)	900

7.8 일본어 관련 수학

7.8.1 일본어로 숫자표기

일본어에서 4, 7, 9는 읽는 방법이 두 가지가 있다. 4는 し가 死자와 발음이 같다는 이유로 よん으로 읽기도 한다. 7은 しち가 死地와 발음이 같기 때문에 なな로 읽기도 한다. 그리고 9의 く는 苦(괴롭다)와 발음이 같기 때문에 きゅう라고 읽기도 한다.

7.8.2 숫자 (기수) 헤아리기

숫자	일본어읽기	일본어쓰기	숫자	일본어읽기	일본어쓰기
0	레이(제로)	れい(ゼロ)	50	고쥬-	ごじゅう
1	이찌	いち	60	로꾸쥬-	ろくじゅう
2	니	に	70	나나쥬-	ななじゅう
3	산	さん	80	하찌쮸-	はちじゅう
4	시(욘)	し(よん)	90	큐-쥬-	きゅうじゅう
5	고	ご	100	햐꾸	ひゃく
6	로꾸	ろく	200	니햐꾸	にひゃく
7	시찌(나나)	しち(なな)	300	삼뱌꾸	さんびゃく
8	하찌	はち	400	욘햐꾸	よんひゃく
9	큐-(쿠)	きゅう(く)	500	고햐꾸	ごひゃく
10	쥬-	じゅう	1000	센	せん

Chapter 7 제2외국어 수학 학습 자료들

숫자	일본어읽기	일본어쓰기	숫자	일본어읽기	일본어쓰기
11	쥬-이찌	じゅういち	2000	니셴	にせん
12	쥬-니	じゅうに	3000	산젠	さんぜん
13	쥬-산	じゅうさん	4000	욘센	よんせん
14	쥬-욘	じゅうよん	5000	고센	ごせん
15	쥬-고	じゅうご	10000	이찌.망	いちまん
16	쥬-로꾸	じゅうろく	20000	니찌.망	にまん
17	쥬-시찌	じゅうしち	1억	이찌.옥	いちおく
18	쥬-하찌	じゅうはち	1/2	니분노.이찌	にぶんのいち
19	쥬-뀨-	じゅうきゅう	1/3	산분노.이찌	さんぶんのいち
20	니-쥬-	にじゅう	3/4	욘분노.산	よんのさん
30	산-쥬-	さんじゅう	0.1	레이노.이찌	れいのいち
40	욘-쥬-	よんじゅう	0.27	레이노.나나로꾸	れいのななろく

7.9 라틴어 약자

라틴어 약자	의미
cf.	(confer) compare 비교하라, ~을 참조하라
e.g. (example given)	(exempli gratia) for example, for instance, 예를 들면
et al.	(et alii) and other people
etc.	(et cetera) and so on, and so forth
i.e. (in essence)	(id est) that is to say, 즉, 다시 말하면
QED	(quod erat demonstrandum) which was to be shown, which was to be demonstrated
vs.	(versus) against
a fortiori	with even stronger reason
a posteriori	from effects to causes, reasoning based on past experience

원서읽기를 위한 수학용어사전

(표 계속)

라틴어 약자	의미
a priori	from causes to effects, conclusions drawn from assumptions
ad hoc	for this purpose, improvised
ad infinitum	never ending
caveat	a caution or warning, 경고
ceteris paribus	all other things being equal
de facto	from the fact
inter alia	among other things
ipso facto	by the fact itself
non sequitur	it does not follow
per capita	per head
prima facie	at first sight
status quo	things as they are
vice versa	the other way round, 거꾸로, 반대로, 역 또한 같다
circa [sə́ːrkə]	대략, ~쯤, ~경 (약자로는 C., ca, cir., circ.를 사용)

기타 수학에 나타나는 약자

QED	quite elegantly done (위의 표 QED 참조)
AWD	And We're Done
W^5	Which Was What We Wanted
OBOB	off-by-one-bug
WLOG	Without Loss Of Generality

Chapter 8 알아두면 유용한 것들

Chapter 8 알아두면 유용한 것들

8.1 영어 문자의 상대도수

다음 표는 영어의 각 문자가 문장에서 나타나는 상대도수를 나타내어 이를 이용하면 행맨(Hangman)게임과 같은 게임이라든지 암호해독 등에 도움이 된다.

영어에서 각 문자가 나타나는 상대도수표			
문자	상대도수	문자	상대도수
e	12.702%	m	2.406%
t	9.056%	w	2.360%
a	8.167%	f	2.228%
o	7.507%	g	2.015%
i	6.966%	y	1.974%
n	6.749%	p	1.929%
s	6.327%	b	1.492%
h	6.094%	v	0.978%
r	5.987%	k	0.772%
d	4.253%	j	0.153%
l	4.025%	x	0.150%
c	2.782%	q	0.095%
u	2.758%	z	0.074%

그룹으로 나눠서 본 영어 문자의 상대도수		
그룹	영어원문에서 나타나는 도수	각 문자의 상대도수 범위
e	12.7%	12% 이상
t,a,o,i,n,s,h,r	56.9%	6% - 9%
d,l	8.3%	4%
c,u,m,w,f,g,y,p,b	19.9%	1.5% - 3%
v,k,j,x,q,z	2.2%	1% 이하

8.2 문장부호

!	Exclamation Point (익스클레메이션 포인트)
¡	역느낌표 : 스페인어에서 느낌문장 처음에 ¡를 사용하고 끝에 !를 사용.
?	interrogation mark, question mark
¿	역물음표 : 스페인어에서 의문문장 처음에 ¿를 사용하고 끝에 ?를 사용.
"	quotation Mark (쿼테이션 마크)
#	crosshatch (크로스해치), number sign
$	dollar sign (달러사인)
%	percent sign (퍼센트사인)
@	at sign (엣 사인, 혹은 엣)
&	ampersand (앰퍼센드) & 는 z 다음에 해당하는 27번째 라틴문자 알파벳으로 여겨졌던 시기도 있었다. ampersand는 간단하게 "and"대신에 사용하는 기호이다.
'	apostrophe (어퍼스트로피)
*	asterisk (아스테리스크)
-	hyphen (하이픈)

Chapter 8 알아두면 유용한 것들

.	period (피리어드)
/	slash (슬래시)
\	back slash (백슬래시)
:	colon (콜론)
;	semicolon (세미콜론)
^	circumflex (서큠플렉스), circumflex accent 위첨자를 나타낼 때 주로 쓰인다. (예) 3^5 은 3^5 (예) 프랑스어의
`	grave (그레이브)
{	left brace (레프트 브레이스)
}	right brace (라이트 브레이스)
[	left bracket (레프트 브라켓)
]	right bracket (라이트 브라켓)
|	vertical bar (버티컬바)
~	tilde (틸드) : 스페인어에서 알파벳 n 위에 ~ 가 붙어 이 된다.
´	저작권 기호 tittle : 영어 소문자 i와 j의 위에 붙인 점을 "tittle" 이라고 한다. (예) not one jot or one tittle : 티끌만큼도, 일점일획이라도 (마태복음 5장 18절)
...	ellipsis(생략부호)는 누락 또는 대화중 정지된 부분을 표현할 때 사용.

참고 (1) A 'jiffy' is an actual unit of time.

: 1초를 100으로 나눈 시간, 즉 0.01초

(2) jiffy 는 잠깐동안, 순간을 의미하며 in a jiffy 는 곧, 즉시라는 뜻이다.

8.3 기타사항

☞　RHS = right-hand side　　우변
　　LHS = left-hand side　　좌변

☞　자정(midnight)과 정오(noon)는 a.m.12 와 p.m.12 이 어느 것도 아니다.

☞　장제법 long division
　　단제법 short division
　　조립제법 synthetic division

다각형의 변

base : 밑변이라고 말하는데 다음과 같은 경우가 있다.
　　(1) 삼각형에서는 높이에 수직인 변을 말함(예각, 둔각 삼각형에서 적용됨)
　　(2) 이등변삼각형에서는 등변이 아닌 길이가 다른 한 변을 말함
　　(3) 사다리꼴에서 평행한 두 변을 말함

side : 변 또는 옆면을 말한다.
　　예를들어 삼각형은 3개의 변을 가지고 있다고 한다.

leg : (1) 직각삼각형에서 빗변을 제외한 직각을 낀 두 개의 변을 말함
　　(2) 이등변삼각형에서 base(밑변)이 아닌 길이가 같은 두 변을 말함
　　(3) 사다리꼴에서 평행한 두변을 제외한 나머지 두변을 말함

Chapter 8 알아두면 유용한 것들

☞ cathetus : 직각삼각형에서 직각을 낀 두변 (leg 라고도 함)을 말하며, 그 두 길이가 다를 때는 minor (또는 shorter) cathetus 는 짧은 변을 말하고, major (또는 longer) cathetus 는 긴 변을 말한다.

참고 복수형은 catheti 이다.

☞ For a right triangle, the term "leg" generally refers to a side other than the one opposite the right angle.
직각삼각형에서 직각을 바라보는 변을 hypotenuse 라 하고 직각에 이웃한 두 변을 leg 라 한다. 이등변삼각형에서 같은 두변을 leg 라 하고, 나머지 변을 base 라 한다. 등변 사다리꼴에서 평행한 위와 아래 두변을 base 라 하고 등변을 leg 라 한다.

☞ 따라서 직각삼각형은 3개의 side(s)를 가지며, 1개의 hypotenuse 과 2개의 leg (또는 cathetus)를 가진다.

☞ 수학의 정의에서 if 는 필요충분조건(if and only if)의 의미이다.

☞ 수학에서 정리는 if 부분(~이면)과 then 부분(~이다)으로 구성되었다. 그리고 if 부분은 여러 개의 조건(또는 가정)으로 구성되었는데 이들의 공통부분(교집합)일 때 then 부분이 성립한다. if 부분의 어떤 조건을 제외할 때 then 부

분이 성립하지 않는 반례(counterexample)를 얻을 수 있다. 반례란 보편적인 명제를 반증하는 예를 말한다.

☞ the number of + 복수명사 + 단수동사
   a number of + 복수명사 + 복수동사

☞ square root 제곱근
   ● The square root of 64 ($\sqrt{64}$) is 8.
   : 64의 제곱근($\sqrt{64}$)은 8이다.
   ● Three is the square root of 9. : 3은 9의 제곱근이다.
   ● The square[second] root of 4.
   : (The root of 4 또는 Root 4 또는 $\sqrt{4}$) is 2.
   : 4의 제곱근은 2이다.

☞ cube root 세제곱근, 입방근(立方根)
   ● The cube root of 64 ($\sqrt[3]{64}$) is 4.
   : 64의 세제곱근은 4이다.
   ● The cube[third] root of 27(i.e. $\sqrt[3]{27}$) is 3. :
   : 27의 세제곱근은 3이다.

   참고 영어로 된 교재에는 cube root를 사용하며 cubic root는 나타나지 않는다. cubic은 cubic centimeter, cubic meter, cubic inch 등에 사용된다.

Chapter 8 알아두면 유용한 것들

1부터 40까지의 제곱과 그 제곱의 제곱근 표			
제곱(square)	제곱근 (square root)	제곱(square)	제곱근 (square root)
$1^2 = 1$	$\sqrt{1} = 1$	$21^2 = 441$	$\sqrt{441} = 21$
$2^2 = 4$	$\sqrt{4} = 2$	$22^2 = 484$	$\sqrt{484} = 22$
$3^2 = 9$	$\sqrt{9} = 3$	$23^2 = 529$	$\sqrt{529} = 23$
$4^2 = 16$	$\sqrt{16} = 4$	$24^2 = 576$	$\sqrt{576} = 24$
$5^2 = 25$	$\sqrt{25} = 5$	$25^2 = 625$	$\sqrt{625} = 25$
$6^2 = 36$	$\sqrt{36} = 6$	$26^2 = 676$	$\sqrt{676} = 26$
$7^2 = 49$	$\sqrt{49} = 7$	$27^2 = 729$	$\sqrt{729} = 27$
$8^2 = 64$	$\sqrt{64} = 8$	$28^2 = 784$	$\sqrt{784} = 28$
$9^2 = 81$	$\sqrt{81} = 9$	$29^2 = 841$	$\sqrt{841} = 29$
$10^2 = 100$	$\sqrt{100} = 10$	$30^2 = 900$	$\sqrt{900} = 30$
$11^2 = 121$	$\sqrt{121} = 11$	$31^2 = 961$	$\sqrt{961} = 31$
$12^2 = 144$	$\sqrt{144} = 12$	$32^2 = 1024$	$\sqrt{1024} = 32$
$13^2 = 169$	$\sqrt{169} = 13$	$33^2 = 1089$	$\sqrt{1089} = 33$
$14^2 = 196$	$\sqrt{196} = 14$	$34^2 = 1156$	$\sqrt{1156} = 34$
$15^2 = 225$	$\sqrt{225} = 15$	$35^2 = 1225$	$\sqrt{1225} = 35$
$16^2 = 256$	$\sqrt{256} = 16$	$36^2 = 1296$	$\sqrt{1296} = 36$
$17^2 = 289$	$\sqrt{289} = 17$	$37^2 = 1369$	$\sqrt{1369} = 37$
$18^2 = 324$	$\sqrt{324} = 18$	$38^2 = 1444$	$\sqrt{1444} = 38$
$19^2 = 361$	$\sqrt{361} = 19$	$39^2 = 1521$	$\sqrt{1521} = 39$
$20^2 = 400$	$\sqrt{400} = 20$	$40^2 = 1600$	$\sqrt{1600} = 40$

☞ perfect square 완전 제곱

: 정수(整數)의 제곱으로 이루어진 수; 1, 4, 9, 25 등

☞ root of unity : 1의 n제곱근(根).

☞ Imaginary numbers are mostly used to represent roots of polynomial equations.

원서읽기를 위한 수학용어사전

: 허수는 주로 다항식의 제곱근을 나타내는데 사용된다.

☞ 각의 종류

acute angle	예각, 0도 보가 크고 90도 보다 작은 각
right angle	직각, 90도
obtuse angle	둔각, 90도 보다 큰각
straight angle	평각, 180도
reflex angle	180도 보다 크고 360도 보다 작은 각
adjacent angles	이웃각, two angles that have the same vertex (corner point) and a common side and don't overlap.
vertical angles	맞꼭지각, two nonadjacent angles formed by two intersecting lines
complementary angles	여각, two angles whose measures add up to 90 degree. One angle is the complement of the other.
supplementary angles	보각, two angles whose measures add up to 180 degree. One angle is the supplement of the other.

☞ internal angle (내각) + external angle (외각) = 180°

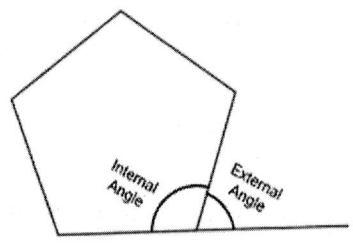

● 두 직선이 다른 한 직선과 만날 때
 (두 직선이 서로 평행이 아니어도 된다.)

Chapter 8 알아두면 유용한 것들

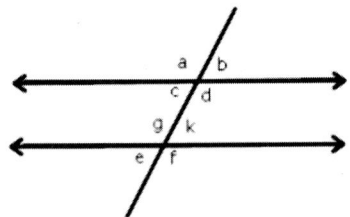

☞ 그림에서 a 와 f 는 alternate exterior angles (외엇각)이다. 역시 b 와 e 도 alternate exterior angles 이다.

☞ 그림에서 c 와 g 는 same-side interior angles (동측내각)이다. 역시 d 와 k 도 same-side interior angles 이다.

☞ 그림에서 c 와 k 는 alternate interior angles (내엇각)이다. 역시 d 와 g 도 alternate interior angles 이다.

☞ 그림에서 a 와 d 는 vertically opposite angles (맞꼭지각)이다. 위 그림에는 여러 쌍의 vertically opposite angles 이 있다.

☞ 그림에서 b 와 k 는 corresponding angles (동위각)이다. 위 그림에는 여러 쌍의 corresponding angles 이 있다.

주의   이들 각은 쌍으로 나타나기 때문에 angle 이 복수형 angles 로 나타난다.

polyominoes

monomino	1가지
domino	1가지
tromino	2가지
tetromino	5가지
pentomino	12가지 : 이 중에 두껑없는 상자를 만들 수 있는 것은 8가지
hexomino	35가지 : 이 중에 접어서 상자를 만들 수 있는 것은 11가지

☞ triangle(삼각형)의 명칭

Triangles are closed figures with three sides

equilateral triangle	등변 삼각형
equiangular triangle	등각 삼각형
isosceles triangle	2등변삼각형
scalene triangle	3변의 길이가 모두 다른 삼각형, 부등변 삼각형
acute triangle	예각 삼각형
right triangle	직각 삼각형
obtuse triangle	둔각 삼각형

☞ quadrilateral (사변형)의 명칭

Quadrilaterals are closed figures with four sides.

square	all sides are the same length, all angles are right angles.
rhombus	all sides are the same length, two pairs of sides are parallel.
rectangle	two pairs of sides are the same length, all angles are right angles.
parallelogram	two pairs of sides have the same length, two pairs of sides are parallel.
trapezoid	one pair of sides is parallel.

Chapter 8 알아두면 유용한 것들

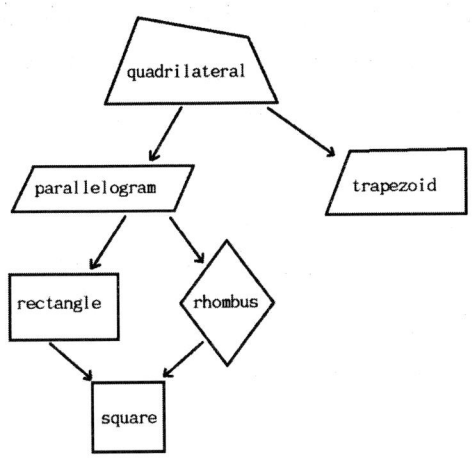

☞ A circle is a plane figure bounded by a curved line. Every point of the curved line is equidistant from the center.

☞ kite [kait] 와 deltoid [déltɔid]
2개의 이웃하는 두 변이 각각 같을 때, kite 는 볼록(convex) 이고 deltoid 는 오목(concave) 이다.

☞ 5가지 정다면체의 이름

정사면체	regular tetrahedron
정육면체	regular hexahedron 또는 cube
정팔면체	regular octahedron
정십이면체	regular dodecahedron
정이십면체	regular icosahedron

☞ 양수(陽數)와 음수(陰數)는 한자 正數(정수), 負數(부수)를 번역한 것으로 우리나라에서는 양수와 음수로 사용하지만 중국과 일본에서는 정수와 부수를 사용한다.

☞ order(계) 와 degree(차)
degree 는 다항식에서 사용하는 것으로 2차 다항식, 3차 방정식 등과 같이 사용된다. order 는 순서를 포함하고 있어서 미적분에서 1계도함수, 2계도함수 등과 같이 사용된다.

☞ 미국 동전이야기
미국 수학교과서에는 미국동전에 관한 내용이 많이 나온다. 1달러를 동전으로 교환하는 방법과 같은 경우의 수도 나타난다. 따라서 자주 등장하는 미국 동전에 대해 소개하기로 한다.
현재 미국에서 통용되는 동전은 penny (1 cent), nickel (5 cent), dime (10 cent), quarter (25 cent), ½ dollar (50 cent), dollar (100 cent)의 6종류가 있다. 과거 미국에서는 다양한 동전이 발행되었다. Half Cent, Two Cents, Three Cents, Half Dime, Twenty Cents, Quarter Eagle($2.50), Half Eagles($5.00), Eagles($10.00), Double Eagles($20.00)등이 현재의 이름을 갖고 있는 동전들과 함께 사용되었다. 미국에서는 Philadelphia, Denver, San Francisco 에 있는 조폐국에서 주로 동전을 발행하고 있다.

Chapter 8 알아두면 유용한 것들

각 동전은 만들어진 곳을 동전의 연도 옆이나 밑에 표시하는 데, 아무표시가 없거나 P가 적혀있으면 Philadelphia에서, S가 적혀있으면 San Fransisco에서, D가 적혀있으면 Denver에서 만든 것이다. 과거에는 S표시 동전이 통용되기도 했으나 현재 S는 기념주화 또는 Proof화에 표시되고 시중에 유통되는 동전은 Philadelphia와 Denver동전이다. 참고로 현재는 금과 은이 합금된 동전은 기념주화와 수집용으로 발행되는 동전 일부분에 들어있다. 최근 미국 동전의 앞뒤 모습이 예전과 달라졌으며 25 cent 동전은 1999년부터 매년 5개 주씩 2008년까지 50개주의 이름과 그 주를 대표하는 상징물을 동전에 넣어 발행하고 있다. 2009년 부터는 미국 준주(state 가 아닌 territory)에 대한 기념 quarter를 발행하고 있다. (더 자세한 사항과 최근 변동사항은 인터넷을 이용하여 "United States Mint"를 검색하기 바람.)

왼쪽부터 차례대로 penny(1 cent), nickel(5 cent), dime(10 cent), quarter(25 cent), half dollar(50 cent)와 dollar 동전. 윗줄은 앞면, 아랫줄은 뒷면 모습이다.

☞ 점화식(漸化式)에서 점화(漸化)란 '각 항이 차례대로 이루어지는'이라는 뜻으로, '불을 붙이는'의 점화(點火)와는 한자로 적으면 확연이 다르다.

☞ negation 부정(否定)과 indefinite 부정(不定)
"이다"의 부정(否定)은 "아니다"이고, 해가 무수히 많이 존재하는 경우에 그 방정식은 부정(不定)방정식이라고 한다.

☞ $a \times b$ 는 정확히 말하면 "b 를 a 에 곱한다."이다. 흔히들 실수로 "a 에 b 를 곱한다."고 말하는 것은 틀린 것이다.

참고문헌

[1] Richard N. Aufmann and Vernon C. Barker, Basic College Mathematics (3rd ed.) Houghton Mifflin, Boston (1987).

[2] Mervin L. Keedy and Marvin L. Bittinger, Introductory Algebra (5th ed.) Addison-Wesley, Reading, Massachusetts (1987).

[3] Richard N. Aufmann and Vernon C. Barker, Intermediate Algebra (2nd ed.) Houghton Mifflin, Boston (1987).

[4] Mervin L. Keedy and Marvin L. Bittinger, College Algebra (4th ed.) Addison-Wesley, Reading, Massachusetts (1986).

[5] 동아 새국어사전, 두산동아(주), 제5판, 2009년.

[6] 시사 엘리트 영한사전, YBM/si-sa, 2013년.

[7] АНГЛО-РУССКИЙ СЛОВАРЬ МАТЕМАТИЧЕСКИХ ТЕРМИНОВ, МОСКВА <<МИР>> 1994. (영어↔러시아어, 수학 용어 사전, 모스크바, 1994년)

[8] Russian-English Vocabulary with a Grammatical Sketch - To be used in reading Mathematical Papers, AMS, 1950.

[9] 안병섭, 러시아어-한국어 사전, 일념, 1987년.

[10] Dictionnaire de Poche Français-Corèen, 민중서관, 1972년.

[11] 동아 프라임 佛韓辭典, 두산동아 1998년.

[12] Portuguese Dictionary
(Portuguese-English, Português-Inglês), Larousse 2007.

[13] Gran Diccionario de Inglés
(inglés-español, español-inglés), Barcelona. 2003.

[14] 丁申寬 & 梁國相 編, Student's English-Chinese & Chinese-English Dictionary, 12th ed, Peiking, 1995.
[15] Deutsch-Koreanische Wörterbuch, 민중서관, 1983년.
[16] The Nystrom Desk Atlas, Nystrom, 2004.
[17] 콘사이스 數學辭典, 創元社, 1977년.
[18] Marcie F. Abramson, Painless Math Word Problems, Barron's 2001.